Further praise for *In Pra*

"A reader-friendly digest of scientific studies. . . . [*In Praise of Walking*] delivers a great deal of salutary, practicable common sense."

—Sam Sacks, *Wall Street Journal*

"*In Praise of Walking* [is] a backstage tour of what happens in our brains while we perambulate. Walking makes us healthier, happier and brainier. . . . [Shane O'Mara] knows this not only through personal experience, but from cold, hard data." —Amy Fleming, *Guardian*

"A fascinating new book that examines the multitudinous benefits of this form of locomotion." —Lydia Slater, *Harper's Bazaar*

"Convincing and compelling. . . . *In Praise of Walking* is peppered with insights about everything from 19th-century poets and flâneurs to modern-day experiments with subjects playing video games in fMRI scanners." —Helen Davies, *Sunday Times*

"Compelling and wise, *In Praise of Walking* points the way to the human adventure." —Richard Louv, author of *Our Wild Calling* and *Last Child in the Woods*

"Full of insights. . . . An accessible and thought-provoking discussion of walking as a key to human success."

—Gina Rippon, author of *The Gendered Brain*

"This eloquent tribute to walking moves seamlessly between neuroscience and literature and is perfectly pitched for nonspecialists who will no doubt be inspired to kick-start or amp up a walking program."

—Cynthia Lee Knight, *Library Journal*

Also by Shane O'Mara

*Why Torture Doesn't Work: The Neuroscience of Interrogation*

*A Brain for Business – A Brain for Life*

# IN PRAISE OF WALKING

*A New Scientific Exploration*

SHANE O'MARA

**W. W. NORTON & COMPANY**
*Independent Publishers Since 1923*

First American Edition 2020
First published by The Random House Group Ltd. under the title
*In Praise of Walking: The New Science of How We Walk
and Why It's Good for Us*


Printed in the United States of America
First published as a Norton paperback 2021



Manufacturing by LSC Communications, Harrisonburg

Library of Congress Cataloging-in-Publication Data

Names: O'Mara, S. M. (Shane M.), author.
Title: In praise of walking : a new scientific exploration / Shane O'Mara.
Description: First American edition. | New York, NY : W. W. Norton &
Company, 2020. | "First published by The Random House Group Ltd. under
the title IN PRAISE OF WALKING: The New Science of How We Walk and Why
It's Good for Us." | Includes bibliographical references and index.
Identifiers: LCCN 2019046247 | ISBN 9780393652086 (hardcover) | ISBN
9780393652093 (epub)
Subjects: LCSH: Walking. | Human locomotion.
Classification: LCC QP310.W3 O53 2020 | DDC 612/.044—dc23
LC record available at https://lccn.loc.gov/2019046247

ISBN 978-0-393-86749-7 pbk.

W. W. Norton & Company, Inc., 500 Fifth Avenue, New York, N.Y. 10110
www.wwnorton.com

W. W. Norton & Company Ltd., 15 Carlisle Street, London W1D 3BS

3 4 5 6 7 8 9 0

# CONTENTS

# INTRODUCTION

What is it that makes us human? What quality is it that makes us different to all other living creatures? Language usually comes top of the list – it is a unique human capacity, without doubt.[1] Other species communicate with each other, and often do so flawlessly and elaborately, with call signs for food and danger, for example. But no other species has anything like human language, with its infinite capacity to carry meaning, content and culture.

We humans also use elaborate tools, and we train other humans in their use; moreover, our tool use evolves over time. But other species do use tools, albeit not with the variety and creativity that we do. Our unique human propensity to cook our food is often cited too: it is true that no other species cooks their food. Cooking gives us nutrients and sustenance from sources we couldn't otherwise use. But this begs the question – how do we gather and transport food for cooking? Another item usually on the list is the exceptional investment we make in our children and adolescents, raising and caring for them for extended periods – a commitment far greater than any other species makes.

There is, however, one entry often omitted from the list – something that is one of our major and singular adaptations (an alteration to our biology that aids our survival), one that is regularly overlooked in the popular mind. It is common to all of the adaptations just mentioned, as well as many more. This is our ability to walk, and especially to walk upright on our two feet, an adaptation known as 'bipedalism', freeing our hands for other tasks.[2] Almost all other land animals are quadrupeds – they walk on four limbs. Walking is a marvellous and seemingly simple feat, and it is a feat that robots have yet to emulate with anything like the fluidity of humans and other animals.[3]

Walking makes our minds mobile in a fashion denied other animals. The (now nearly forgotten) neurologist and phenomenologist Erwin Straus captured the sense of how intertwined walking is with our identity and experience in 1952, commenting that our 'upright posture is an indispensable condition of man's self-preservation. Upright we are, and we experience ourselves in this specific relation to the world.'[4] Our upright posture changes our relationship to the world, including, as we shall see, our social world.

Our close relatives, the chimpanzees, use an intermediate form of bipedalism, where they walk using a combination of their hands and feet. This adaption is known as 'knuckle-walking', and it is not an especially efficient way of getting around.[5] Many birds also walk on land on two legs – but they do not walk with an upright spine.[6] Their spinal columns are not perpendicular to the ground with a mobile head on top. For humans, walking bipedally has involved dramatic modifications and adaptations across the whole length of our bodies, from the top of our head all the way to our toes.

What does walking on our two feet give us that makes us different? In evolutionary terms, bipedalism enabled us to walk out of Africa and to spread all over the world – to the far distant glaciers of Alaska, and the sun-baked deserts of Australia. It's a unique ability that has defined human history.

Walking upright has also given us all sorts of other physical advantages. Bipedal walking frees our hands, meaning we are able to carry food, weapons and children. Shifting locomotion to our feet, stabilising our balance along the spine and hips, has allowed us to throw stones and spears, to creep along and attack others with primitive stone axes, to gather up the stolen spoils of assault and combat, and to quietly disappear into the night. We have been able to carry our young – often over great distances – by simply putting one leg in front of the other. Walking upright makes our minds mobile, and our mobile brains have marched to the far horizons of our planet.

But the benefits of walking aren't solely confined to our evolutionary history – walking is hugely beneficial for our minds, our bodies and our communities. Walking is holistic: every aspect of it aids every aspect of one's being. Walking provides us with a multisensory reading of the world in all its shapes, forms, sounds and feelings, for it uses the brain in multiple ways. Walking together can be one of the best experiences of walking. Social walking – marching in concert and with purpose – can be an effective goad for real change in society. Walking is so vitally, centrally, important to us, at both individual and collective levels, that it should be reflected in the way we organise our lives and societies. Our public policymaking needs to fully embrace why walking makes us so distinctively human, and should feed in to our urban and suburban planning. I look forward to the day when physicians the world over write prescriptions for walking as a core treatment for improving our individual and aggregate health and well-being. In fact,

GPs in the Shetland Islands have already started prescribing beach walks as a preventative treatment against maladies of brain and body.[7]

Throughout this book, we will celebrate the full sweep of human walking, from its origins deep in time, through how the brain and body perform the mechanical magic of walking, to understanding how walking can set our thoughts free, all the way to the most social aspects of walking – whether it's a four-ball in golf, a country ramble, or a march to try and change society. Along the way, we will draw the lessons to be learned and show the benefits for the individual and society. And these lessons are manifold, and easy to apply.

I will show how walking makes us social, by freeing our hands for tools, and for gestures – movements that allow us to signal meaning to others. Walking allows us to hold hands, sending out signals of exclusive romantic involvement; walking allows us to provide physical support to each other; marching in protest is a common feature of our free political lives, which is why the prevention of assembly and of marches is one of the first orders of an autocrat. Walking is good for the body, good for the brain, and good for society at large.

But the converse is also true. We pay the price for our lack of movement, whether it arises because of the environments we occupy, the design of our offices, or from just being idle and sedentary. I want to show in this book how imperative it is that we start walking again. Our brains and our bodies will be the better for it; our moods, clarity of thought, our creativity, our connectedness to our social, urban and natural worlds will all be the better for it. It is the simple, doable, personal fix we all need.

The emerging science is giving us a clear picture: regular walking confers enduring and substantial benefits on individuals, and on society at large. This book celebrates both the science of walking and the unalloyed joy of going for a good walk. I want to place an ostensibly simple behavioural change into a central position as the driver of positive psychological and physical well-being. It is an activity that almost everyone can engage in, and that comes naturally to us. Our brains and bodies are built for movement in our daily lives, in our natural and built environments: regular movement improves our thinking, feeling and creative selves in myriad ways, as well as improving our health.

It's time to get up and walk our way to a better life – go see the world as it is, and as only we humans can.

# 1.
# WHY WALKING IS GOOD FOR YOU

We overlook at our peril the gains to be made from walking, for our health, for our mood, for our clarity of mind. Many of us live now in a deeply unnatural environment, where we spend long periods of the day sitting with our eyes focused on screens, perhaps a half-metre or so from our eyes. When we stand up, and then walk around and move about, our posture changes, with our torso and spinal column shifting to a single vertical axis from our head down through our back, and, through our legs and feet, contacting the ground. By contrast, when we sit, the weight of our body trunk is largely concentrated on the lower back, and in particular, on the coccyx, that little collection of bones that comprises our vestigial human tail.[1] The coccyx anchors a remarkable lattice of tendons and muscles extending across the spine and down the upper legs in particular, the gluteus muscles of the upper thighs, which are vital for walking. Little wonder that lower-back pain is one of the most common ailments in the developed world.

How silly, then, that the remedy – to regularly get up out of your seat and walk about – is so little understood or practised. Long periods of immobility also cause changes in muscle: fatty deposits build up in leg muscle, and, as we age, we lose muscle mass in part because of our immobility ('sarcopenia'). There are many other changes too: our blood pressure changes, as does our metabolic rate (the rate at which we burn energy). But when we stand up, things suddenly change in brain and body: we become 'cognitively mobile', our minds are in movement, our heads swivel, our eyes dart about. Our brain activity changes when we move about, with electrical brain rhythms that were previously quiescent now engaged and active. We become more alert, our breathing changes, and our brains and bodies are readied for action. The French philosopher Jean-Jacques Rousseau commented that 'I can only meditate when I am walking. When I stop, I cease to think; my mind only works with my legs.'[2]

Here's a walking memory of mine: I'm at a student conference in Belfast during the dreary and seemingly never-ending 1980s. I take a long walk up the Malone Road, past Queen's University into the centre of town. I pass through the numerous security cordons. Young soldiers with serious weapons are patrolling the city, looking in shopping bags for bombs and guns, talking nervously to each other in English accents. There's plenty of tension in the air. The Loyalist politician Ian Paisley's campaigning against the Anglo-Irish Agreement is a constant backdrop, as are the terrible atrocities, the many bombings and murders. The city is alive, though. A city is hard to kill.

When I cast my mind back over this walk on my first visit to Belfast, I remember that I walked past the much-bombed Europa Hotel. I then walked east toward Botanic Avenue, and then took a long loop back around the streets and roads to the rear of the Europa Hotel. Why this route? Just because I could; that's what being on foot does for you. It's early Saturday afternoon, the weather is grey, and there's a hint of rain in the air. Wandering about, I accidentally find myself walking on Sandy Row, the Loyalist epicentre of Belfast. The murals are amazing, and a little frightening, to someone from the sedate and peaceful south. I walk quickly on, eventually connecting with the Lisburn Road, and finally find my way back to where we students were all staying on the Malone Road. Here, in Belfast, a walk is a walk into a past that is still present; as the old maxim has it, 'the past hasn't even passed'.

Wrapped up in this little personal journey are many of the elements of the hidden story of walking: mental time-travel to recall details, reminiscences about a walk, orientation and successful navigation through an alien urban environment, the little frisson of fear that still comes to me when I remember the security cordons and the murals. We now know that the brain systems relating to all of these functions are in constant communication and support each other's functioning. And, crucially, these brain systems are not perfect. My memory has tricked me a little. It has simplified the route, and left out significant details. I remember Botanic Avenue as being almost opposite the Europa Hotel. It's not, as a look at a map tells me. Botanic Avenue is at an acute angle that runs on to Great Victoria Street, which is actually where the Europa Hotel is. And, weirdly, I have excised most of the detail about the relative location of Sandy Row and the Europa Hotel. I remember

Sandy Row as more or less directly behind the Europa. It's not: Sandy Row is further south than that. I am left to imperfectly recall the gists of locations, places, things; I do not possess a faithful video recording somewhere in my brain of the route I took all those years ago.

This is the key point underlying our episodic and event memories: they are imperfect, gist-like, extracting meaning, focusing on certain salient points, and ignoring others.[3] There is more information out in the environment than our mobile minds can capture, and more than we need to know. How we move, what we look at, who we talk to, what we feel as we move: these are central components of our experiences. They might enter into our recall and be laid down as traces in our brains. We are not disembodied brains travelling through space and time: we feel the ground beneath our feet, the rain on our face; perhaps peering into the unknown, but in doing so we are extending our range of experiences of this complicated world. And all the while we are silently creating memories of where we have been, and making maps of the world we have experienced.

It's possible to demonstrate the brain-changing power of simply getting up and walking about. A straightforward experiment called the 'Stroop' task – devised by American psychologist John Ridley Stroop[4] – is used to test 'cognitive control', in other words, the ease or otherwise with which you can direct and control your attention and thinking. The Stroop task is a colour-and-word identification task with a twist. Participants are presented with lists of colour names (red, green, blue, black, etc.). These are printed either in the same colour (for example, the word 'red' printed in red) or in another colour (the word 'red' printed in green). Participants are asked to, as quickly as they can, name the colour of the printed word. Typically, when the printed word and the colour it names are congruent, response times are rapid and accurate. By contrast, when the printed word and the colour it names are incongruent, response times are much slower.

Often, Stroop-task performance is impaired under dual-tasking conditions. For example, a participant might be asked to engage in the colour naming, while simultaneously monitoring sentences played through earphones, and listening out for a particular word or phrase which they must identify by pressing a button. The Stroop effect is very reliable and easy to detect; it is often explained as requiring the paying of selective attention to certain aspects of the visual stimulus, while actively suppressing attention to other (automatic, attention-grabbing, prepotent) aspects of the visual stimulus, and then selecting and making the appropriate response.

But what happens if you add movement into the mix? The experimental psychologist David Rosenbaum and his colleagues at Tel Aviv University wondered if merely standing up might have an effect on Stroop performance.[5] They found, over a series of three experiments, that when a participant is standing up, the Stroop effect for incongruent stimuli – where performance should be slower – is, in fact, faster than is normal compared to when they are seated. It is as if the mere act of standing mobilises cognitive and neural resources that would otherwise remain quiescent. Moreover, recent studies show that walking increases blood flow through the brain, and does so in a way that offsets the effects of sitting around.[6] Regularly interrupting prolonged bouts of immobility through the simple act of standing up changes the state of the brain by calling on greater neurocognitive resources, constituting a call to action as well as a call to cognition.

As well as improved cognitive control, it's clear that walking confers many, many other benefits. We all know that it is good for our heart. But walking is also beneficial for the rest of our body. Walking helps protect and repair organs that have been subject to stresses and strains. It is good for the gut, assisting the passage of food through the intestines.[7] Regular walking also acts as a brake on the ageing of our brains, and can, in an important sense, reverse it. Recent experiments asked elderly adults to participate in thrice-weekly, and relatively undemanding, walking groups.[8] In the regular walking group, over the course of a year, the normal ageing of the brain areas providing the scaffolding for learning and memory is somewhat reversed in the walkers, by perhaps about two years or so. An increase in the volume of these brain areas was also found; this is quite remarkable in itself, suggesting that the act of regular walking mobilises plastic changes in the very structure of the brain, strengthening it in ways similar to how muscles are strengthened when worked.

One way of interpreting the literature on ageing and walking is straightforward: you don't get old until you stop walking, and you don't stop walking because you're old. Lots of regular walking, especially if conducted at a high tempo, with an appropriate rhythm, forestalls many of the bad things that come with ageing. Walking is also associated with improved creativity, improved mood, and the general sharpening of our thinking. Periods of aerobic exercise after learning can actually enhance and improve recall of the previously learned material. Reliable, regular, aerobic exercise can actually produce new cells in the hippocampus, the part of the brain that supports learning and memory. Regular exercise also stimulates the production of an important molecule that assists in brain plasticity (known as brain-derived neurotrophic factor, or BDNF).[9] The phrase 'movement is medicine' is correct: no drug has all of these positive effects. And drugs often come with side effects. Movement doesn't.

When walking in the beautiful Glendalough valley once, I felt the thrumming of many running feet. I stopped, and was treated to the sight of four or five red deer running through the glen. It was late autumn, breeding season, and I could hear the stag roaring and calling. This is something else walking does for you: you see, smell and feel things as they are, not through a windscreen at speed. Walking allows you to confront the personal, instead of insulating you from it. Like many people, I also drive, and I always take the train to work. But walking is special to me as a form of transport. Walking allows me to walk it off, whatever *it* is. Walking clears my mind, allowing me to think things through. Natural movement brings with it experiences and demands on the body and brain that do not arise from other types of movement. Cars, bicycles, trains and buses all divorce you in different ways from the environment, you are mechanically propelled, sometimes insulated behind glass, travelling too fast, worried about crashing, trying to find that new song on the radio. There is a peculiar passivity to it: you are sitting, yet you are moving at speed. This can never be true of walking: one foot must go in front of the other until you get there, under your own steam. You make your own way, and experience the world close at hand, at your own speed, in your own way.

But, how do we know that walking has all these multifarious benefits for our minds, bodies and quality of life? What's the evidence? The evidence is extensive and, as we will see in the course of this book, shows that walking enhances every aspect of our being, from our physical health, to our mental health, to our social lives and beyond.

*

This may seem an obvious point, but when we're walking our brains are in motion too. In fact, as we shall see, we evolved as a mobile species: we walk about, we move, we seek new sources of information from the world. In other words, we are not just brains locked in a skull, we are minds in motion – we are 'cognitively mobile'. The study of how we think, how we reason, how we remember, how we read, how we write, is known as the study of cognition. Typically, the scientific investigation of cognition occurs in a laboratory, using carefully controlled experiments and a range of methods and tests that measure cognitive abilities.

Almost anything that moves in a reliable and consistent way can probably be measured somehow. The movements made can be various and manifold. They might be the pattern of eye movements that a person makes: where they look, and for how long, at particular locations on the screen can be captured; the rapid flickering of increases and decreases in pupil size might be measured; the electrical responses of the brain might be analysed; reaction times might be assayed; how much the person fidgets in the experimental chair ascertained. And, in the latest generation of experiments, participants might perform these complex tasks while lying in a brain-scanning machine, which uses a variety of advanced methods to measure and to localise activity in the brain associated with the performance of a particular cognitive task.

There are two principal methods of brain imaging. The first and by far the most popular is magnetic resonance imaging (MRI), which comes in two principal flavours: functional (thus, fMRI) and structural (sMRI). MRI is a medically safe, non-invasive procedure that allows you to (in principle) see the brain at work with details down to the millimetre. The other major brain-imaging tool is positron emission tomography (PET), which involves the injection of radioactive tracers into the blood, and mapping their uptake in differing brain regions during differing tasks. PET is a technique with comparatively poor spatial localisation, compared with MRI, and is a little unpleasant, especially if you have a needle phobia. PET has found specialised uses particularly in the development of new drug treatments for brain and other disorders. MRI, by contrast, does not involve any injections, and offers much greater localisation in terms of structure and function. MRI and PET have allowed us an unprecedented view of the brain at work – and especially of the human brain at work.[10]

Let's now imagine you are asked to participate in an fMRI experiment. You are placed on the bed of the MRI scanner, and slowly inserted into the bore at the centre of the machine. First up, an sMRI: a picture of multiple slices through your brain, to check that there are no abnormalities or other problems present. Assuming this goes smoothly, you will then be instructed in the task you will perform in the fMRI. Here, you will first gaze at a small cross on a screen (this is called eye fixation), and then you will be asked to perform a task. Keeping to the theme of this book, this task might be a spatial navigation task. You might have a joystick, and you have to find your way around a complex three-dimensional maze. We can predict, based on what we know from experiments on rats and on humans, that we will see a very high degree of activity in the hippocampal formation, as well as activity in brain regions involved in motor movements. How do we show there is activity in the hippocampal formation, specific to the task, and not to other aspects of the task? Here is where control experiments are absolutely essential. Often, a subtractive logic is employed: task-irrelevant activity is subtracted from the task of interest. You might ask the subject to move the joystick according to a verbal instruction, but not while exploring the maze, so that they are engaged largely in visuo–motor behaviour.

This laboratory-based, experimental approach has been remarkably powerful. It has allowed us to test and extend the standard model of human cognition. However, it does come with certain limitations. The particular limitation we concern ourselves here with is our ability to measure what it is that goes on in the brain while the brain is moving around, when the mind is mobile 'in the wild', as it were. The experimental psychologist Simon Ladouce and his colleagues at Stirling University argue (correctly, in my view) that our understanding of cognition has progressed more slowly than it could or should have done, because past and current generations of psychologists and neuroscientists have not studied mobile minds and brains with the intensity that perhaps we might have done.[11] To be fair to legions of experimenters, this has occurred of course because putting the lab into the wild is difficult. Studying the actions of the mind in motion can be done, but it is not easy. Realistically, to study cognition in the wild requires taking what is best of laboratory practice and somehow making lab instruments mobile so that we can measure what it is that people think, say and do while they are walking about.

The latest generation of mobile technologies are becoming well known to us all and these can be adapted and used to capture behaviour while we are out and about. Many, if not most, of us now have smartphones. These usually now come equipped with apps to measure the number of steps, speed of walking, our diet and many other things besides. Expanding on these and other technologies allows us to capture more of what the brain is doing when cognition is mobile. Smartphones have proven to be particularly useful. Participants can be pinged at different times of the day and they can be asked what they are doing, how they are feeling and what it is they are planning to do, among other questions. This is known as 'experience sampling'.[12]

While there are indirect ways of studying how walking changes the brain, specifying and understanding the underlying mechanisms can be more difficult. Relating these changes to the activity of these brain cells, circuits and systems to overall cognition and behaviour is more difficult still. However, we do now have the beginnings of an understanding of how walking affects activity in the brain. In turn, we are now starting to understand how walking changes the brain in order to prepare for action.

Imagine for a moment being a cat, sitting and waiting for your prey. There's a rat nearby which, in turn, is moving about looking for something tasty to eat. Imagine being that cat, stealthily stalking the rat. Your visual system is more sharply attuned simply because you are moving quietly about. You pick up information more quickly; your paws are readied to capture your prey.

Now imagine being the rat heading back to a burrow or nest. It is semi-dark, and cat-you and rat-you are both operating at the edges of your visual acuity. You can, perhaps, each smell the odour trail of the other, but the odour trail is, perhaps, indistinct and does not provide a reliable path to track down the prey, or escape from the predator. Unless your place of refuge is completely secure, the better escape strategy is to move quietly and carefully, relying on the extra tuning that your visual system has when walking. Similarly, moving around, moving your head, your eyes, allows you a better chance at picking out your prey – the meal that you so badly need tonight.

Here, we have an interesting 'evolutionary arms race'. Activity in the visual areas of the brains of both the rat as prey, and the cat as predator, is sharpened and tuned by walking.[13] Walking allows you to capture your prey more easily, but equally walking allows you to escape the predator more easily. There are two competing mobile cognitive systems, one cat-based, one rat-based, each tuned to defeating the aims of the other. And the activity of each system is sharpened by the same thing: walking. This leads us to an important, and general, conclusion: walking markedly changes activity in the brain in subtle, important and powerful ways.

This cat–rat, predator–prey example allows us to consider mobile cognition by thinking through what happens to activity in brain cells, circuits, systems and then behaviour. What happens to your sense of how you see things when you are walking? Does walking affect seeing? How fast can I see something when I'm walking and paying attention, compared to when I am sitting and paying attention? Walking changes activity in the parts of the brain that are concerned with seeing, and it changes them in a variety of positive ways, designed to make responding to what is happening in the real world quicker and more effective.

Let's think for a moment about how cognition might be affected by movement. We can think of the brain (somewhat simplistically, it has to be said) in the following way: the brain receives inputs from the outside world (the sensory side of the nervous system), and processes these in some way (the central component of the nervous system). In turn, the results of this processing can affect behaviour through some form of output (the motor side). Activity in these differing component parts can then be measured during walking. And the picture that emerges is that walking measurably changes brain activity for the better. Hearing, sight and reaction times all improve during active movement.

Of course, we're not simply layabouts; and our mobility presents particular problems for data collection. As we shall see, conducting experiments with mobile rats and mice is now relatively straightforward. Experiments in mobile humans, however, require somewhat more ingenuity.

*

The Via Alpina, or Alpine Way, courses over eight countries (Austria, France, Germany, Italy, Liechtenstein, Monaco, Slovenia and Switzerland), and consists of five lengthy interlocking trails. These trails amount to about 5,000 kilometres. The Via Alpina trails are truly ancient in origin, and occasionally, archaeological finds on the trails tell interesting and disturbing stories of their past. In one such case from 1991, a 5,000-year-old mummy of a middle-aged male was discovered on what is now the Austrian–Italian border. He was given the name of Ötzi the Iceman.[14]

Poor Ötzi had come to an unpleasant ending: X-rays of his body showed that he had been struck in the left shoulder by a deeply embedded flint arrowhead, and had then been hit hard on the head. His arms show some defensive wounds. It's not entirely clear which wound killed Ötzi. Neither the arrowhead nor the strike to his head are likely to have resulted in instantaneous death. He may simply have died as a result of blood loss from his shoulder wound. Almost inevitably, it seems, in death Ötzi would suffer further indignities – he's become quite the celebrity on the Via Alpina, with local restaurants now serving Ötzi pizza, and Ötzi ice cream.[15]

How does Ötzi compare with a modern human? Ancient humans would very often have been nomadic, unlike their more sedentary twenty-first-century brothers and sisters. How would this nomadic lifestyle have affected Ötzi's body?

A 2011 experiment was able to roll back the years, allowing us to see what his body might have looked like and how it would have changed because of this nomadic lifestyle. Researchers studied how a sixty-two-year-old, reasonably active male adapted and responded to walking the long trail across the Alps. The unnamed Italian man walked 1,300 kilometres of the Via Alpina over the course of three months.[16] Before starting, he reported to the laboratory where he was measured from tip to toe. Measures were taken of all of the critical aspects of his bodily functioning: his breathing capacity, his muscle strength, how lean his body was, the various components of his blood, and a whole variety of other measures. He was then equipped with a mobile physiological laboratory that he carried with him. This laboratory in miniature consisted of instruments in a rucksack, along with tools to allow him to repeatedly take and measure his own blood chemistry. The portable devices allowed the researchers to develop a picture of our modern Ötzi's ongoing adaptation to his prolonged mountainous trek.

And the good news here is that it is never too late for anyone to start walking, even over long distances. Despite being reasonably fit, our modern-day Ötzi had never taken a journey on foot of this length and duration before, yet the measurements showed that his body adapted quickly and easily to the rigours of the journey, including overcoming the effects of mild oxygen starvation at altitude. (The Via Alpina varies between 0 and 3,000 metres above sea level; altitude sickness usually happens above 1,500 metres, where oxygen levels drop to 84% of the oxygen available at sea level. By 3,000 metres, oxygen levels drop to 71% of the oxygen available at sea level.)

There were positive changes in virtually every single measured area of his functioning. His body mass index – often used to determine obesity – declined by about 10%. On the other hand, his percentage of measured body fat fell dramatically, by about a quarter in total, as a result of the continued exercise of the journey. (Need to lose weight? Don't go to the gym; go for a really, really long walk. And do it in nature, over a period of days to weeks. It will be far more beneficial to you.)

Overall, our modern-day Ötzi walked just over 1,300 kilometres in sixty-eight days – about 19 kilometres a day – although there was considerable variation in the amount he walked each day: on some days he only managed about five or six kilometres, whilst on other days he clocked up to forty-one kilometres. This variation, of course, reflects the difficulty of the underlying Alpine terrain. Five kilometres walked while ascending perhaps 2,000 metres over a rocky mountainside trail is a considerable achievement indeed, whereas walking forty kilometres on a gentle, well-maintained, downward trail over a seven- or eight-hour period is perhaps, relatively speaking, not such an onerous walk. The length our modern Ötzi walked is comparable to other endurance walks that other humans have engaged in. It's this ability to walk substantial distances over difficult terrain, in relatively short periods of time, by making steady, reliable, daily progress that was key to our species' long journey out of Africa, as we'll see in Chapter 3.

One especially important benefit to our modern-day Ötzi was a major and sustained decrease in the kinds of fats (triglycerides) thought to underlie at least some forms of heart and cardiovascular disease. His extended walk resulted in an approximately 75% decrease of these triglycerides. There was also a large increase in the production of fats (high-density lipoproteins, of the sort found in olive oil and fish oil) believed to protect the heart. So, here we have strong evidence from a deep case study of a late-middle-aged male that an exercise regime based on daily walking can markedly protect the heart, not just by making the heart fitter (although it certainly does this), but by reducing factors in the bloodstream that can cause heart disease. We can conclude from this case study that, even in late middle age, the body, motivated appropriately by its brain, can adapt dramatically and for the better as a result of an extended period of daily endurance walking.

Are these bodily changes an anomaly, or do they reflect some underlying general set of processes, common to us all? Our Italian participant's inflammatory and other markers of disease also fell dramatically across every single one measured. Does he have some physiological peculiarity, or some odd genomic idiosyncrasy responsive to such extended periods of walking?

Testing very remote, unconnected human populations is one way to rule out such concerns. A study conducted in Bolivian Amazonian hunter-gatherers suggests that they too are similar to our modern-day Ötzi (or, rather, that he responds to activity in the same way they do). The evolutionary anthropologist

Hillard Kaplan and colleagues from the University of New Mexico conducted a study on 705 members of the Tsimane who live in the Amazonian jungle.[17] These are hunter-gatherers who live principally on a diet of fish, wild game and high-fibre carbohydrates. They have a very low dietary intake of low-density lipoprotein, or LDL. Smoking is vanishingly rare among the Tsimane (although they are subject to high levels of parasites). The Tsimane are very active during the course of the day, occupied by farming, hunting, food preparation, household chores and parenting. They walk everywhere and do not used wheeled transport, or ride animals.

Remarkably, Kaplan and his colleagues found that almost all Tsimane tested have markers of cardiac health that are better than the healthiest Western societies. Coronary artery calcium (CAC) scores are a measure of the extent to which there are calcium plaques (with a build-up of fats and other materials and detritus) in arteries. These plaques can solidify and even eventually block the free flow of blood through an artery, causing heart attacks and strokes. The CAC scores of the Tsimane are a fifth or less of those of Western populations, and fully 65% of them have a CAC score of zero. So an eighty-year-old Tsimane actually possesses the vascular age of an American in their mid-fifties.

Although the study did not include actigraphy (direct measurements of movement), observational data collected suggests that the Tsimane spend many hours per day engaged in physical activity – multiples of that found in a Western, sedentary, industrialised population. They spend time food gathering over extended distances, hunting, fishing, foraging and other activities necessary for survival. We can safely and reasonably conclude that high levels of activity (principally walking) can, along with dietary changes, markedly help protect the heart against factors that promote heart disease. Moreover, these factors can be reversed both by activity (for the better) and inactivity (for the worse). Modern-day Ötzi shows that these malign changes can be reversed quickly, by walking, and walking lots, whereas a sedentary lifestyle worsens them.

Sadly, no records were taken of our modern-day Ötzi's mood. I am willing to bet, though, that we would see two differing components of his mood. His moment-to-moment mood would have reflected the ongoing challenges of such a walk: being wet, too hot, too cold, or hungry and thirsty; being frustrated with the inconveniences of a nomadic life (where do I sleep? Where do I eat? Where can I go to the loo?). But his long-term sense of well-being, as well as clarity of thinking or indeed clarity of consciousness, will have moved upwards, perhaps enduringly so.[18]

*

Whilst our Italian guinea pig was outfitted with a mobile laboratory to measure his functioning, modern-day technology gives a new and more convenient solution for studying mobility and activity in the population at large, as mentioned earlier: the smartphone. Smartphone apps can passively track our walking steps and walking routes. We can log our age, weight, track our heart-rate and even our blood-oxygen levels. We can measure our relative activity levels in ways that were unimaginable a decade ago.

I always try to keep track of the number of steps I take every day, using my smartphone. I have a target that I attempt to hit as a minimum per day, though I'll often try to exceed it.

Why do I use a smartphone? The simplest reason is that it is impossible, without smartphone or other pedometer data, to actually track your walking steps reliably and consistently. It is difficult to remember how much walking you do from one day to the next. And it is next to impossible to know your average walking speed from hour to hour and day to day accurately without some form of pedometer. Having the data to hand allows you to calibrate how much you actually walk against what you think you walk; there is no way to gild the lily on this. Self-reporting is subject to all sorts of individual failures and problems.[19] We are reasonably good at providing relatively accurate statements about how we are currently feeling, but reporting on how we felt seventy–two hours ago is a different matter entirely. Capturing walking data reliably and consistently allows you to have a really fine-grained picture of how much walking you do, and when you do it, over the course of a day. A smartphone really can be your personal, individual lab in your pocket.

Smartphone penetration has increased dramatically in all societies and across all income groups, in both the developed and developing worlds. This almost total smartphone penetration now allows the capture of both individual-level walking data and country-level walking data, with high precision, over long periods of time. This is perfectly illustrated by the size of data sets now available; for example, computer scientist Tim Althoff and colleagues at Stanford University created a data set consisting of 68 million walking days captured from 111 countries, involving 717,527 people, aged from their mid-teens to their mid-seventies.[20] This is what the phrase 'big data' actually means: a vast lake of data, made of up almost uncountable drops of data gathered from hundreds of thousands of people.

After performing a variety of data checks and data cleanups, they ended up with a final data set comprising 66 million walking days from 46 countries, involving 693,806 people. They were also able to capture age, body mass index (BMI), and gender, and went on to use this data set to build up a picture, at the country level, of patterns of activity regarding human walking. Additionally, they were able to use another data set from the US, which analyses – across a variety of measures – the walkability (or otherwise) of 69 US cities, from across the whole of the US. Importantly, as we will see in Chapter 5, some of these cities are from the same geographic locale, are similarly affluent, and have similar demographics, but show great differences in levels of walking, depending on the walkability of the city. They found huge variations within and between countries for the number of steps taken by individuals in each country. The best predictor of obesity, as measured by body-mass index (BMI), is not the absolute number of steps taken by individuals within any country, but rather the inequality in the number of steps taken within a country between males and females.

All sorts of interesting details came out of this study. For example, males walk more than females at all ages, from the mid-teenage years through to the seventies. On the other hand, females on average have lower BMIs than males. BMI, however, increases on average for lower activity levels. The average person in their analysis took 4,961 steps per day (though there was considerable variation behind this average, from above 14,000 steps to just a few hundred steps per day). There was also considerable country-to-country variation. In Japan, for example, the average number of steps taken per day was 5,846, whereas in Saudi Arabia, this figure was 3,103. The gender gap in steps walked was apparent across all countries tested, and in large part this gender gap gave rise to a substantial fraction of the observed differences in BMI.

What use do I make of my own personal, pocket-dwelling, laboratory? My smartphone walking app is a goad to action, no doubt about it. The step counter is my guilty conscience. I always want to hit at least 9,500 steps per day, which my phone records, but I prefer getting above 12,000 steps per day and am really happy with more than 14,000 steps per day. I now tend to make the 9,500-step target almost every day in each month, and hit 12,000 steps about eighteen days per month, and 14,500 steps about ten days per month. There is no way that I could remember with any consistency or accuracy the number of steps I take per day, without using pedometer data from my smartphone. My self-report data would be utterly unreliable. And this is fine: we should offload these dull tasks to pocket-held robots.

Why have I chosen these step targets? Well, in part because the smartphone app that I use allows me to measure my own performance against those of the population of people using the app more generally. Knowing how important activity is for health, I think it reasonable in my own case to ensure that I am at least consistently in the top 20% of the population that use my particular smartphone and its native app. Is this a reasonable target? Well, without knowing the population level data that Althoff and colleagues have captured, I might be subject to peculiar biases captured just by my smartphone app. It appears not, and it appears further that my own step targets are reasonable (at least compared to others).

*

Beyond health, walking brings many other benefits for brain, body and behaviour, which we'll go on to explore throughout this book. We'll also discuss the many poets and writers who have written eloquently on the wonders of walking as a spur to mood, creativity and thinking. Writers are amongst the best at recognising walking's essential and intrinsic virtues and rewards. The poet I return to, time and again, is T. S. Eliot. I find Eliot's poetry has a cadence and rhythm that are remarkable, especially if read aloud. His great modernist poem 'The Love Song of J. Alfred Prufrock' (1915) is a journey on foot, and a journey through states of mind. It is a poem set to the rhythm of a long urban walk, undertaken uncertainly as evening falls. The opening lines are an invitation to walk the city:

> Let us go then, you and I,
> When the evening is spread out against the sky
> Like a patient etherized upon a table;
> Let us go, through certain half-deserted streets,
> The muttering retreats
> Of restless nights in one-night cheap hotels
> And sawdust restaurants with oyster-shells

Eliot extends an invitation to an unspeaking, and unseen, other to walk with him in the late evening, to explore on foot the low areas of the city. The journey on foot is essential to the cadences of the poem. Eliot does not ask the unseen other to cycle with him, to take a cab with him, to travel in the train with him. It is an invitation to walk.

Walking promises types of experience denied other forms of transport, no matter how attractive they might be. There will be the sights, conversations, the sounds of others, the smells of the lonely, window-ledge pipe-smokers who will be seen at 'dusk through narrow streets'. The feeling is one of Eliot having a conversation with himself whilst walking. There is a blurring of interior and exterior worlds, while Eliot, as Prufrock, walks uncertainly along. The poem is a notable, though oblique, tribute to walking and wandering in the urban and social worlds; the walking tempo carries you along through an imagined evening.

As Eliot says, 'Let us go', and explore now the wonder that is walking: all of it. The science, the history, the complex interactions of bone and muscle and nerves, stumbling, ambling, mooching, wandering, traipsing, strolling, tramping, striding, stepping. Our journey will take us from ancient Africa, to the mechanics of movement, into the innermost recesses of the brain as it maps the globe, to walking in concert and with purpose in order to change the world.

# 2.
# WALKING OUT OF AFRICA

Let's start our walking story with a simple, sea-dwelling organism: the sea squirt. There's an oddity in the life cycle of the sea squirt.[1] This is a creature that lives in rock pools, and which during its early life has a single and basic eye connected to a simple brain which is connected to a simple spinal cord, rendering it, in effect, a small, water-dwelling, vertebrate cyclops. The fact that it has a spinal cord places it in the same group of animals as us humans, cats, fish, birds and other members of phylum Chordata, as it is technically known. During its larval, or early, developing, immature life, it is free-floating, and can swim, hunt and maintain homeostasis (the ability of an organism to keep itself at a particular set point in order to stay alive, like a thermostat keeps heat at a set point in a home). When the larval sea squirt is hungry, it hunts, returning itself to a nutritional homeostasis.

It can even maintain balance while swimming. Keeping its soft underbelly to the bottom gives greater protection from predators, so it knows which way is up or down. The larval sea squirt keeps its balance using a statocyst – a hollow ball of cells (a sac), lined with hairs connected via nerves to its brain. Inside the sac, there is what looks like a small marble, a mineralised mass known as a statolith. The statolith, obeying the laws of gravity, rolls to the bottom of the sac, a bit like a marble inside an everted tennis ball. This position signals that the larval squirt is the right way up, with its stomach to the bottom. If the statolith rolls away from this steadying position, the hairs it touches signal that the larval squirt has moved away from an upright position, and it will right itself.

Eventually, as it grows, the squirt transitions to a fixed stage, sticking itself to a convenient rock. There it consumes its own semi-brain, spinal cord and eye, none of which it now needs. It does this by simply reabsorbing these cells and using them as a meal. The sea squirt becomes little more than a stomach with some reproductive organs attached.[2] It is a vegetative grabber of whatever food particles happen to tickle its fronds. Now that it is no longer moving, it no longer needs a 'brain'.

Sea anemones display a similar, though even simpler, life cycle: they go from being free-moving polyps to cementing themselves to a rock, absorbing the nerve clumps they possess that function like a basic, if distributed, brain.[3] The reverse pattern is true of certain species of jellyfish: as immature polyps, they are attached to rocks, and their fronds float in the water.[4] As they grow they become free-floating, and start to develop a complicated nerve net that allows them to engage in patterned movements, to attack prey and to ingest food.

The larger lesson is clear: brains have evolved for movement. If you're going to be stuck, unmoving, in one place, with your food all around you, then why do you need a costly brain?

Trees don't have brains, nor do animals with a sessile (non-moving) life. But if you're on the move, on the prowl for food or a mate, or looking for shelter, you need a brain. The wider world is complex, and presents you with lots of problems to solve, and to solve quickly. You need a brain to control directed movement towards possible sources of food and shelter and mates, and away from things that might eat you. That brain must also learn to recognise good, safe, edible food and then direct the movements of the body to catch or harvest it.

For the motile sea squirt, movement allows it to occupy a particular ecological niche, a niche it vacates when it becomes sessile. The same is true of the anemone; it ceases to move, and vacates an ecological niche. By contrast, jellyfish reverse this journey, leaving a sessile lifestyle in favour of a motile one. We humans make a similar journey: we start off helpless and almost immobile at birth, but develop the skills to crawl, and eventually gain autonomy through walking. For sea-dwellers, swimming is movement at its most elemental, providing them with the reading of their ecological niche needed to survive.

Getting around, moving from place to place, somehow, is one problem you must solve if you are a motile species. One method is to drift wherever the ocean currents take you. Or perhaps you're active: do you swim along in the water with fins, wriggle, slither and snake along the seabed or, perhaps, push yourself against the ground using limbs? Differing solutions arise deep in our evolutionary past, in the seas and oceans of the world. The larval sea squirt uses one method for getting around, with propulsion provided by rhythmic movements of its tail fin. Another method for moving is walking: using extensions of the body which make contact against a reasonably solid substrate, against which you can push yourself.

These methods of movement in water and on land make use of a combination of soft tissues, muscles and the like, attached to bones. Soft tissues typically do not fossilise, and are therefore usually absent from the fossil record. The historic approach to understanding our evolutionary past has been through the reconstruction of fossils and inferring likely patterns of relationship and divergences between differing species. This approach has its limitations, not the least of which arises because of the loss of the tissue attached to the bones. Figuring out the likely functions of soft tissues in their absence is very difficult indeed.

Occasionally, though, other types of fossils are present: 'trace fossils' providing an imprint of the past, like footprints in sand. Some are apparent through commonalities in the genes we share with other species. Walking connects us to our deep, evolutionary past; occasionally, we are lucky enough to glimpse the fossil remains of the walk itself. This is often a walk that occurred at a 'deep time' so long ago that it is far beyond our limited human experience, though not beyond our comprehension. This is time measured in tens or hundreds of millennia, not years or decades.

A walking female figures strongly in the reconstruction of our deep evolutionary past. She has been called 'Walking Eve' for the light she has shed on our evolutionary history. Hers are probably the first human footprints to be fossilized and to surface from deep time. They are 'trace fossils' providing clues about height, weight and gait, as well as foot shape (morphology), allowing us to compare these ancient footprints to modern ones. These are what remain of a young woman who walked across the edge of Langebaan Lagoon, a saltwater lagoon in what is now the Western Cape province of South

Africa, adjacent to the Atlantic Ocean, some 117,000 years ago.[5] This location is not so hugely different now to how it might have appeared then. 'Walking Eve' would still recognise the position of the stars, as there is no light pollution; the sand underfoot would feel much the same. She might be familiar with the blue waters, sandy shores and distant mountains. She left just three footprints in mud, which filled with dry sand blown into them through the happenstance of a sandstorm, preserving her footprints for us to wonder at, generations later.

New ways of understanding our evolutionary past have also become possible through modern genomic technologies, which, when combined with detailed anatomical and fossil studies, present a likely timescale for how and when the genes controlling modern mammalian walking arose. When T. S. Eliot wrote in 'Prufrock' that 'I should have been a pair of ragged claws / Scuttling across the floors of silent seas', he was closer to the truth than he could possibly have realised. The science of genetics has revealed that the gene complex governing walking, via the control of the development and patterning of muscle, sinew, nerve and bone, was already present in the deepest reaches of evolutionary time. One lesson is very clear from these studies: evolution is an astonishingly conservative engineer of adaptation when it hits upon solutions that work.

Walking (whether on land or on the seabed) involves extending muscles followed by flexing muscles in alternating sequences, a rhythmic pattern controlled by nerve cells of the spinal cord. In mammals, walking involves a combination of extensor and flexor muscles to extend and contract limbs. A not dissimilar arrangement occurs in fish fins, allowing swimming. Does this surface commonality hint at a deeper, perhaps genetic, relationship? It's now becoming clear that evolution has provided both fish and mammals with an underlying genetic programme for movement, but one that is adapted to the form of movement of these differing animals. The programme is similar in both species, with circuits controlled at the level of the spinal cord, giving rise to biological 'central pattern generators' which generate rhythmic muscle movements.

Evolution by natural selection suggests there should be deep genetic links between all species.[6] These genetic links should and do extend to the control of movement and walking, as we shall see. Analysis of the genes of widely disparate species confirms this indeed is the case. Let's first take a simple example of these commonalities: the workings of an individual cell.[7] At a simple functional level, the internal economy of all the cells that constitute our bodies and the cells of other species – from worms to fish to monkeys – must solve very similar problems: they must keep invaders (sources of infection) out; they must allow nutrients in; they must remove waste; maintain a stable fluid balance; and all of the myriad other things required to stay alive. Nor must these cells become cancerous: these cells must stop dividing and die when appropriate. A so-called 'minimal genome', widely shared across cell types, as well as across species, supports and sustains these essential cellular functions.[8]

What we can say with some certainty is that the genetic networks that control locomotion also appear to be highly conserved across differing species. The surface appearance (or phenotype) of one animal relative to another might appear quite different, but in terms of the underlying genetic architecture (genotype), they are substantially similar. We can test this idea by examining the genes controlling similar functions in very disparate species, and look at the degree of similarity or difference between them. Are the genes that support the circuits controlling walking similar in all land-dwelling animals, for example? That's easy to test, as all land-dwelling animals derive from a common tetrapod (four-legged) ancestor – and the answer is yes.[9] What might be more interesting is to test commonalities between species that live and move in very different environments: on land and in the sea.

The genes that control the expression of these circuits are known as the Hox genes. Hox genes have the ultimate function of controlling the sequence in which body segments appear during embryological development. These genes are so alike in differing species that the Hox genes from a chicken can be placed into the developing fly (while simultaneously removing the fly Hox genes) and the fly develops perfectly normally – even though the last common ancestor for these two species existed more than 600 million years ago.[10]

In a pioneering study, developmental biologists have examined the gene networks that underlie fin and limb motor neurons controlling movement in the fish *Leucoraja erinacea*, commonly known as the 'little skate' (a cold-blooded sea-dweller with bilateral fins), and in the mouse (obviously, a warm-blooded, land-dwelling, four-limbed mammal).[11] Both skate and mouse have to solve at least some similar problems, how to move about being a particularly important one. Both animals have spinal cords, and muscle and bone patterning along their spinal cords that are symmetrical. The mouse obviously lacks fins, and the skate is lacking in mouse-like legs, but both are able to move around smoothly and easily. The mouse walks about on land, and the skate uses its hind fins to walk along the seabed. The skate, in common with the mouse (and indeed, humans), shows left/right alternation of movement, as well as the pattern of extension and flexion required for walking. Thus, there might just be some underlying genetic programme controlling movement in common between the two species. Legs and fins mounted out of spinal cords must come in symmetrical pairs, in order for them to be of any use for swimming or walking.

This research shows that these genes emerged at least 420 million years ago – the time at which the last common ancestor of the mouse and the skate existed. The gene network supporting the expression of these circuits is common to both the mouse, which has paired limbs, and the skate, with paired fins. The gene network is so complete that it encodes for the expression of the relevant muscles, as well as the nerves that radiate from the spinal cord, to innervate these muscles. In turn, the necessary pattern of repeated reciprocal excitation and inhibition is supported and performed to support fin movements, and paired leg movements allow seabed walking and land walking, respectively. These circuits are, as the jargon has it, 'conserved' across species separated by deep evolutionary time. From a walking skate to a four-limbed, walking, water-margin-dwelling tetrapod to a walking mouse to a walking human: the same gene networks bind us together in a common web, across almost unimaginable periods of time.

So genetic programmes required for walking are present in two very diverse species. Moreover, the skate and the mouse share a common genomic programme supporting movement of the body through a common molecular mechanism – the Hox gene regulatory programme. There are some differences in expression between the species, of course, because fins have a different muscle size and shape, compared to the muscles found in the limbs of mice, although the number of muscles, and the arrangement of the relevant spinal cord bones controlled by the Hox network, are not dissimilar. The conclusion of this study suggests, straightforwardly, that the gene networks that control limb movement – whether walking on the ocean floor, or walking on land – are common to all animals that walk using paired limbs (legs or fins). It follows, therefore, that the gene networks necessary to allow walking on land actually appeared first in water, and before the land was colonised by tetrapods. The surprising, wonderful and counter-intuitive conclusion is that the genes for walking appear to have evolved largely underwater, rather than emerging as fish adapted to life at the edges of water systems, where selection might have favoured fins that found traction in waterlogged mud.

The broader lesson from evolution, therefore, is this: selection and modification of pre-existing adaptations, with the preservation of their key components, in terms of structural morphology, appears to be the rule. Not only is evolution, as Richard Dawkins has it, a blind watchmaker, evolution is a conservative watchmaker, keeping what works for lengths of time that are difficult to imagine, but which are certainly comprehensible, calculable and measurable, reusing the same recipe time and again in one species after another.[12]

There's another marvellous and wonderful walking tale from our deep past that ties this all together, this one from an animal that lived some 380 million years ago. Tetrapods were the first four-legged vertebrates in existence.[13] Their hindlimbs are given the technical term 'peds', and their forelimbs are given the technical term 'mans'. A tetrapod trackway is a trace fossil trackway preserving the walking pattern of a tetrapod, in the fashion that trace fossil human footprints have been preserved at Langebaan Lagoon (and other sites throughout the world). Tetrapod trackways are very rare. One such trackway was discovered by the Swiss geological student, Iwan Stössel, on Valentia Island off the coast of Kerry, Ireland in 1992.[14] These footprints are the oldest, most extensive and longest series anywhere in the world – and they're in Ireland, so of course I had to go and see them for myself.

The tetrapod trackway is in a beautiful setting, overlooking the small islands known as the Skelligs (the largest of which, Skellig Michael, was used as the hideaway of Luke Skywalker in two recent *Star Wars* movies). The day we visited, the Atlantic was in full swell, with waves breaking across the rocks. A simple steel cable fence encircles the footprints and the site is marked with a sign bearing a drawing of a tetrapod. My daughter commented that the drawing looked like a cross between an iguana and a duck-billed platypus with a vertical, rather than horizontal tail. Its forelimbs are certainly smaller than its hindlimbs, and its walking pattern would have been one of alternate flexion and extension of the hindlimbs and forelimbs, so that the animal can move rapidly forward.

The tracks are quite a sight; they continue for several metres in many directions. The hypothesis is that these footprints are of a tetrapod walking along a drying-out riverbed, some 380 million years ago. A variety of calculations show that the creature was likely to have been approximately one metre or so in length. The path inscribed by the tetrapod gives little clue as to its likely activity when its footprints were preserved. We can only speculate if it was ambling happily around in the sunshine, or was quietly tracking prey. Nonetheless, looking upon footprints from this deep period of geological time is chastening, and brings with it the sense of connectedness of all species on the planet. In particular, the commonality obvious between it, as a vertebrate, and other land-dwelling, walking, ambling, rambling, shuffling, running vertebrates, is palpable. Imagining that a descendant might at some point arise which would be bipedal is not that difficult a stretch of the imagination, especially as it is clear now that the genes needed to control our locomotion are present in our deepest common ancestors.

*

We humans, rightly and correctly, display a fascination with our origins. We love to know who our relatives are, we long to know about our ancestors, we also readily anthropomorphise and identify with our close cousins, in particular chimpanzees, orangutans, and other non-human primates. We are very closely related to chimpanzees, and we recognise this, for example, through laws that prevent chimpanzee experimentation and hunting. It is obvious from studying their anatomy and ours that there are very close relationships in human and chimpanzee body structure and function, just as there are variations and differences too. The human genome and the chimpanzee genome differ by less than a few per cent at the deepest genetic level.

But there are of course also many significant differences between both species. Among the primate family, humans are unique in showing 'obligate' bipedalism. We transition from a stable, crawling stance to an upright walking stance in a period of mere months, early in life, and we maintain this stance even in the face of injury. Injuries that paralyse us and make us incapable of controlling our own locomotion are correctly recognised as limiting the normal activities of daily life (hence, the much-needed recognition of wheelchair accessibility for the mobility-impaired). Environmental design policies can make lives better for the mobility-impaired: those in wheelchairs or on crutches; those who use walking sticks or have prosthetic limbs; or those with neurological impairments. Such policies ensure that everyone can participate to the fullest in our mobile societies, adding greatly to the sum of dignity and freedom for all.

Humans present a unique form of animal bipedalism. We stand upright, with our hands free and mobile, and our spinal columns more or less vertical, relative to the ground under our feet. Our lineage, in evolutionary terms, is exceedingly complicated and somewhat murky, although answers to significant questions regarding the evolution of human bipedalism have started to emerge at a rapid rate over the past few decades. Walking connects us to our evolutionary journey – the journey to walking upright with our hands free, heads aloft, and perpendicular spines.

None of our extant primate relatives walk upright as we do. The great apes – chimpanzees, orangutans, and others – most certainly do not, although they can and do raise themselves onto their hindlimbs for the purposes of display, maintaining vigilance or food-sourcing.[15] Moreover, they typically run and walk using all four limbs. This is a less effective and more energy-demanding method of getting around than being bipedal. Humans can cover approximately twice the same distance for the same calorie burn as a chimpanzee, and therefore the range over which humans can roam is much greater for a given amount of food. A different way of saying this is that human bipedalism is much more economical than are the forms of locomotion that our nearest primate relatives adopt. Our close primate cousins are, of course, very suited to the ecological niche that they occupy. They are, in many species, extremely capable climbers of trees, and are able to seek fruit at the limits of tree branches that we humans cannot easily access. Similarly, their typical forward-leaning posture may allow them to be extremely good ground foragers, where they can obtain fallen fruits, nuts, and the occasional small animal for nourishment.

We occupy an intermediate niche: we can ground-forage, to some extent, and we can also collect fruit from trees. We are, though, supremely adapted to the hunting of other species, especially through our adaptation known as persistence hunting, where we can run animals to ground, with a preference for hunting herbivorous (plant-eating) prey. Our other bipedal adaptation is running: we are good, moderate speed, long(ish)-distance runners, especially during conditions of heat. Running generates heat, demands energy, and the runner also requires lots of liquid for eventual cooling through perspiration. We are relatively hairless (facilitating heat loss), and this, combined with walking, allows us to traverse large distances at a relatively low cost in terms of heat generation and energy use. To cool down from the heat they produce through running, most other species must eventually rest, gulp in air, and breathe it out rapidly to cool down (horses are one exception – they can perspire and pant). This puts them at risk of predation by humans: when lying down attempting to cool off, they can be killed, butchered and carried back to camp for roasting over an open fire: something our close non-human primate relatives definitively do not do.

We humans are also superb gatherers of food: we can pick up and carry tubers, fruits, nuts and burgers with grace and ease, and do so over long distances. We can also eat while walking: the relative position of our throats allows food to travel down to our stomachs while we wander along, with our gaze to the far horizons – an adaptation few other animals have. Indeed, as we shall see, there is also an important evolutionary 'trace fossil' visible in the balance we typically strike between being active and conserving energy.

Running and walking are also closely related. Humans are not especially fast runners – we can easily be outrun over short(ish) distances by lots of other species (think tigers and gazelles) – but we are exceptional walkers, possibly the best walkers of all species. And this has been the secret underlying our far-flung dispersion across the face of the earth. We humans are the most dispersed of all animal species, living in the northern and southern extremes of our planet, and at virtually every land point in between. Walking allowed us to probe and extend the edges of our world, and then undertake occasionally risky journeys by boat, travelling to the next island – which we then explored on foot.

Furthermore, we first colonised the planet by walking in small migratory groups – at its core, our walking is social. An African proverb reflecting this reality says: 'If you want to go fast, go alone. If you want to go far, go together.'[16] At a slow, five kilometres a day walking – at most a few hours walking for a family group – for 300 days, you can travel 1,500 kilometres. Do that over a few years, and you will cover many thousands of kilometres. The longest land-only distance on the earth is from the west coast of Liberia on the Atlantic seaboard of Africa to the eastern Pacific coast of China, a distance of some 13,589 kilometres – about a nine-year amble. Bring it up to twenty kilometres per day for 300 days, and it's a little over two years of walking. Over the course of a few generations, humans can more or less get anywhere with a walkable land-bridge or where they can traverse small water courses, perhaps even island-hopping in primitive boats.

*Homo sapiens* undertook the move out of Africa perhaps as recently as 60,000 years ago. There is some evidence, gathered in Israel and other locations, of an earlier dispersal or dispersals, but given the lack of complete fossil evidence, suggestions of multiple African dispersals and intermingling of differing historical human founder populations are being actively investigated.[17] There is little doubt, for example, but that modern humans interbred with the Neanderthals, as our genome carries their ancient signature.[18] Modern genomic technologies, allied to the techniques of fossil hunting and anatomical reconstruction, will likely give more exacting and precise answers over the coming years.

The human fossil record is extremely varied, and is very certainly substantially incomplete. It is not known where bipedalism fully and finally emerged in the depths of our evolutionary history, though it is certainly a feature of fossils gathered in Africa. Fossil specimens collected in Chad, Kenya and other African countries from 6–7 million years ago seem to be bipedal. A fossil of an early hominid collected in Ethiopia, known as Ardi (*Ardipithecus ramidus*), seems also to be bipedal.[19] This fossil is about 1.3 metres in height. The very famous female fossil known as Lucy dates from approximately 3.1–3.2 million years ago.[20] Lucy was an australopithecine, and possessed a very similar anatomy to that of humans, in particular with respect to the shape of her pelvis. *Homo sapiens* emerged as a distinct species perhaps as recently as 300,000 years ago.

The human form of bipedal walking requires changes across the long axis of the body from our head, through the neck, spinal column, pelvis, legs and feet. These differing components must be (approximately) vertically positioned for efficient bipedal walking. When thinking about the positioning of the brain and skull, it is useful to think also about the positioning and orientation of the pelvis. The brain, skull and pelvis are connected via the spinal cord, which is in turn protected by the vertebrae of the spinal column. Changes in the relative position of the spinal cord, so that it is more upright in orientation, must also involve coordinated changes in the position of the skull and the pelvis. At the base of the skull is a large, approximately circular opening – the foramen magnum – through which the spinal cord passes en route to innervating the body, into its final position within the matrix of pelvic bones. Its approximately vertical orientation allows us to stand upright, perpendicular to the ground. In other species, the foramen magnum occupies a more rearward or posterior position. In four-legged animals with spinal cords that are parallel to the ground, the foramen magnum is almost at the very rear of the skull.

The relative position of the foramen magnum has shifted considerably in primates, and most especially so in humans. Without this repositioning of the foramen magnum, an upright posture would be more or less impossible, or, at least, difficult to maintain for long periods of time. And the routine bipedality of us humans would be impossible. Thus, standing upright on two feet and walking involves changes in the relative positioning of bones at the top and the bottom end of the spinal cord: at the top, to allow the spinal cord to emerge from the base of the brain; and at the bottom, to allow the spinal cord to sit in the appropriate vertical position into the lattice of pelvic bones. Of course, this picture implies two things: a consistency of development and selection of particular body plans over time; but more deeply, a variation in the patterning of Hox genes that allows this modification to appear simultaneously across the long axis of the spinal cord. The selection pressures that gave rise to bipedalism in humans continue to be a matter of debate and investigation – these pressures include resources available within the environment, the physical demands of that environment (such as the presence of water courses, treelines, open grasslands, mountains, valleys), and other factors (such as the type of predators present, or the relative abundance of prey). These pressures have driven selection of the necessary gene coding for the differing aspects of bipedalism.[21]

Human bipedality differs from the few other species that are occasionally bipedal (some species or rats and mice, birds and marsupials), because, quickly and early, humans develop bipedalism from a crawling stance. In consequence, the human foot and ankle have evolved to support our form of bipedalism. We are not arboreal, so do not need a side-placed big toe to assist with tree-climbing and the like. Many species, of course, are transiently bipedal; bears and the great apes will often rear up on their hind legs for purposes of display, and in readiness for either attack or defence. Other animals will stand to retrieve food, or as meerkats and some other species do, adopt a hindward stance in order to keep watch for predators. These stances are neither habitual nor are they obligate; the bipedal stance results from the challenge the animal is experiencing at that moment. Walking has also allowed us to capture and gradually to adapt other species to our needs: cattle, dogs, sheep, horses, camels, for example. Cattle, in particular, show a pattern of co-adaptation over time that clearly mirrors human wants (while we, in turn, evolved lactose tolerance): they now produce much more milk and meat than their ancient ancestors would have done.[22]

Bipedalism involves changes across our body axis, from the shape and positioning of our head all the way down to the relative length and position of our toes. Our feet have changed in shape and our ankles now bear much heavier loads. Recent scientific investigations suggest that over many thousands of years the human foot changed from being prehensile (having hand-like properties with a large toe placed in a similar position to the thumb) to having adaptations that favour long-distance walking and running (with the toe to the front, and an arch that favours forward, elastic movement).[23] Walking Eve's footprints at Langebaan Lagoon appear to have a similar form of bipedality compared to that of modern humans.[24] The evolutionary biologist and anthropologist Herman Pontzer makes the point that the earliest hominins were bipedal, but had some apelike features, from their physiognomy (their facial appearance) to the formation of their hindlimbs.[25] This view is entirely consistent with the current available fossil record, but precise dating of the first appearance of modern human feet awaits more fossil finds.

A signal difference between humans and chimpanzees is that humans have a greater range on foot than do chimpanzees and, indeed, *all* our other primate relatives. This ability to range over wide distances suggests that there are changes in how we process energy during walking, and how we conserve energy during rest. The question, of course, is how did this adaptation arise, and what use did we make of this range as we were gradually evolving it? It would be a mistake to think of range and bipedality as being independent of each other. As our transition to bipedalism advanced, so too our range increased, and in turn that amplified evolutionary selection in favour of bipedality. This greater walking range offers greater possibilities for food sourcing, for seeking shelter, and allows an energy surplus available that can be devoted to other things, most obviously allowing an extension of childhood, post-birth development of a very large brain, and reproduction.

The changes in gene expression that allowed us to become bipedal have, over aeons, allowed us to develop a unique pattern of dispersion from our origins in the Great Rift Valley in East Africa. We initially radiated out all over the great continent of Africa, then out across the Eurasian land mass, and eventually dispersed to the Americas and the Asia–Pacific region and thence to Australasia. And we may have taken this journey multiple times, meeting and interbreeding with our ancestors.

*

Our evolutionary journey has marked us in many ways. We are adapted to walk upright, and we are superb gatherers of food, because of our free, mobile hands. Our evolutionary legacy strikes a fine balance between how much activity we engage in and how we conserve energy. Humans, like all other animals, move about the world in a way that attempts to conserve energy. This is a significant evolved compromise: getting sources of energy (food) uses stored energy. Balancing energy expenditure, energy conservation and energy capture – minimising the effort to find and harvest food – has an important evolved adaptive value that now has maladaptive consequences in the modern world. We strike an often precarious balance between expending energy through movement (mostly by walking about), and conserving energy by sitting around or lying about. The relationship between activity, food consumption and weight is complicated, because the body has many feedback mechanisms to ensure that body weight, once attained, remains stable.

Hunter-gatherers (such as the Tsimane) walk everywhere, and often walk very long distances, daily, gathering food and water, while holding weapons, tools and small children. One way to approach the question of what the natural – in other words, evolved – relationship is between our body weights and activity levels is to try and see what the relationship is between these differing factors in humans who still have a hunter-gatherer lifestyle, with all of the walking that implies. There are relatively few hunter-gatherers left in reasonably accessible parts of the world. Examining the lives of contemporary hunter-gatherers might allow us a window into what are for most of us 'ancestral forms' of activity. Few of us now walk long distances to stalk and kill our food; nor do we spend hours daily foraging for roots and tubers, or searching for drinkable water.

One such contemporary group of hunter-gatherers are the Hadza from northern Tanzania, who have been studied by evolutionary biologists.[26] Researchers have examined the relationship between walking and other measures of activity, and weight, in this hunter-gatherer population, comparing them to participants drawn from North American and European demographic groups. Data from hunter-gatherers allows you to investigate what has changed in recent decades. Have our diets changed, or have our activity levels? Or have both changed?

The Hadza live more or less the life of the traditional hunter-gatherer. The males tend to persistence-hunt large live prey with spears and bows and arrows, and the females tend to gather fruits, berries, tubers and honey. In this study, thirty Hadza research participants wore a GPS tracking system, and their height, weight, and daily calories, consumed and expended, were measured. Some findings should not be surprising. Hadza males walked an average of approximately eleven kilometres per day, and Hadza females walked an average of approximately six kilometres per day. Their percentage body fat was about 60% of Western levels, and no Hadza participant tested was obese. By tracking daily expenditure of energy, when weight and body fat of the participants were accounted for, the total energy expenditure of the Hadza was similar to the Westerners. This was true whether measured by activity levels, and also if males and females were examined separately. This finding was contrary to expectations: it was assumed the average Hadza would expend much more energy during the course of the day than would an average Westerner. Instead, they consume less energy (i.e. they eat less).

Moving about in the world, whether by walking, cycling, running, swimming, or whatever, takes energy. This requires that energy be captured (through eating and digesting), and this energy to be stored and burned, as required. Humans are lazy in the very particular sense of minimising the energy burned for any given amount of movement. Human bipedal walking generally and automatically minimises energy expenditure while maximising terrain covered. Experientially, it certainly feels as though we like to converge on walking speeds that maximise the amount of movement that we can engage in for a given amount of expenditure of energy. Experimentally, how can we prove that this is so?

One method is to fit humans with leg exoskeletons: mechanical frames with settings that make walking either easier or more difficult. How we adapt to the different settings of the exoskeletons allows a test of whether or not humans adapt and settle down into particular walking gaits which minimise the amount of energy that they have to expend in order to walk. If the exoskeleton aids walking, we might quickly adopt a gait where we don't try especially hard to walk; we will let the machine do the work. Equally, if the exoskeleton resists our walking, then we might try to walk with great effort, but settle down into the least amount of effort possible to continue walking.

The neuro-engineer Jessica Selinger and her colleagues at Stanford University have adopted just such an approach.[27] They fitted human participants with jointed, leg-mounted exoskeletons, which allowed them to systematically challenge the most efficient method that humans might adopt to walk. The subjects also wore oxygen masks, thereby allowing measurement of oxygen uptake and throughput (known as $VO_2$ max). The exoskeletons were designed to provide differing degrees of resistance to the leg, depending on the step frequency of the participant. The participants walked on treadmills, the speed of which could also be varied.

Participants rapidly and quickly adapted to the forced changes to their gait patterns, typically in a matter of minutes and, once adjusted to their new optimal gait, they could readjust very quickly back to a previous gait which optimised energy burn (as measured by their $VO_2$ max). The general lesson is that humans readily adopt a more economical gait, one that facilitates optimal stepping, and they do so in ways likely to save the greatest amount of energy expenditure. These adaptations are very rapid – far faster than would be predicted from changes in blood oxygen levels, or other internal sensing such as changes in muscle. Independently, these changes are known to be slow. Instead it appears to be the case that humans actively make predictions based on peripheral sensory inputs, and use these predictions to directly modulate walking.

Although wearing an assistive robot exoskeleton causes you to minimise energy expenditure, there is a more general issue: you go to the gym, running great distances on the treadmill, and then as a reward you relax on the couch at home feeling great that you have exercised, unaware that your overall activity levels are lower than they might have been had you not visited the gym at all. Effectively, in evolutionary terms, your body is making you relax after engaging in persistence hunting. This exercise-induced inactivity confounds our standard thinking, which basically suggests that calories in, plus exercise, plus bodily housekeeping, should equal calories out. If the calories in are greater than are needed for exercise and housekeeping, then we put on weight.

The important lesson from these and other studies is this: simply increasing exercise levels alone is not the solution to obesity, for we have evolved behavioural and physiological mechanisms which compensate for increases in activity by driving down subsequent activity as the result of exertion itself. Getting energy expenditure up will not necessarily lead to substantial and sustained weight loss. What is needed is a full accounting of energy intake and use: we need to know how the body balances energy intake, energy storage (i.e. fat deposition) and energy output. We humans are a highly omnivorous species. We scavenge, we hunt, and we prepare a unique range of foods. Being flexible about food sources and being able to source and prepare food in a variety of differing ways confers many adaptive advantages. You can eat where you find food (you scavenge, or you visit your local restaurant); you can eat while walking (having visited your local fast-food joint); or you can bring food back to your place of shelter to prepare it in differing ways (by carrying, with your fellow hunters, some unlucky beast you have hunted and killed, or by going to your local supermarket for the dazzling variety of foodstuffs that appear at the end of a very complex logistics supply chain).

High-fat, high-sugar, energy-dense foods are easily available in Western market economies, but are not easily available to the Hadza. Increasing activity levels to match those of the Hazda, therefore, is not by any means the straightforward and simple solution to the obesity epidemic besetting much of the world. Rather, a change in the type, quality and quantity of calories we consume has to be a major target of public policy, if the epidemic of problems associated with obesity is to be tamed.[28] To be clear: I am not arguing here against activity, or arguing against a general need to increase activity levels. It is clearly and palpably the case that being active is better for every single organ system of the body than being inactive. And better again that this activity is undertaken in large and regular doses, day by day, week by week and year by year, right throughout life. Activity is central and vital for the control of obesity, but it is only part of the picture. Energy intake is also a major element.

Is regular activity a form of easily self-administered medicine, because movement is intrinsically good for us? The evolutionary biologist Daniel Lieberman of Harvard University offers an important evolutionary perspective on our odd relationship to physical activity: it's good for us, yet we tend to avoid it, in order to conserve energy, for the simple reason that's how we evolved. Until recently physical activity was obligatory, and food sources were scarce. As Lieberman puts it, 'humans evolved to be adapted for regular moderate amounts of endurance, physical activity, into late age . . . because energy from food was limited, humans were also selected to avoid unnecessary exertion'.[29]

Consider the following example: many national institutes of health and diet recommend an average daily diet of perhaps up to 2,000 calories for adult females and 2,500 calories for adult males.[30] Let's call that 2,400 for ease: that implies the burn-rate of those calories is 100 calories per hour, on average – that's through both waking and sleeping hours. Those consumed calories must serve all the needs of the body: breathing, walking, thinking, digesting, secreting and so on. The food itself must be digested, nutrients transported around the body from the roots of the hair on the top of your head all the way to the soles of your feet. Digesting the food you eat costs energy (calories) – high-fibre foods with a low energy density are more costly to digest than are high-calorie foods. Your body must produce heat to maintain life (you are warm-blooded, after all), and must actively and passively transport out bodily waste (e.g. carbon dioxide from breathing must be exhaled, food waste from digestion must be voided, and excess liquid excreted through sweat and urine). Taking in more calories than needed means the excess energy must be stored somewhere – and it is, as fat.

Except in cases like the Hadza or the Tsimane, we don't spend many hours wandering about looking for our daily bread.

Smartphone tracking data shows we now engage in a limited range of movement: a few thousand steps per day, on average. We could and should do so, so much more than these paltry few steps. There is clearly not a straightforward relationship between calories consumed, exercise and body weight, but what is clear is that activity is at the core of overcoming the health problems caused by our sedentary lifestyles – and one of the core takeaways from this book is to get out and walk lots over the course of the day, and to do this every day. It will repay you in more ways than you know.

# 3.
# HOW TO WALK: THE MECHANICS

It's the most natural thing in the world. But how, in fact, do we put one foot in front of the other so reliably, regularly, rhythmically?

The decision to walk, run, duck, flinch – to move – originates in the brain. It is a top-down process resulting from some urgent external signal ('there's danger – move and hide'), or some internal signals ('I've eaten too much – I need to go for a walk'). In the case of walking, top-down input from the brain's decision-making and motor movement systems and circuits provide the instruction to move, and signal the type of movement required. Then the pattern-generating local circuit in the spinal cord – whose job is to produce a rhythmic output – takes care of the instructions from there. Think of it like the signal sent by your car keys: these fire up the engine but don't participate in the movement of the car, or the workings of the engine, until the time comes to shut it down again.

These are a complicated set of mechanisms to coordinate, and they require extensive training, and lots of practice. In consequence, the average human child spends a long period crawling, attempting to become upright and balanced, and begins independent walking at around eleven or twelve months of age. Learning to walk is correctly regarded as a significant milestone in individual autonomy; it is also a significant milestone in brain development.

You don't come into the world ready to walk. A newborn foal will stumble up on their feet quickly, because they are preternatural walkers, compared with us humans. They need to be. A newborn gazelle is easy prey for a hungry and lazy lion, unless it can move, and move quickly. In evolutionary terms the ones who didn't, couldn't move, walk, stumble, then run, all became easy meals for lazy, hungry lions. The ones that could move at birth did not, by and large, get eaten, and passed on their advantageous genes. And the evolutionary cog turns another notch.

We humans, however, come equipped with the propensity, but not the capacity, to walk. We have to learn through experience to sculpt the wondrous circuits that allow us to walk for a lifetime. Learning to walk – like learning to talk, or to see, or to hear – is one of those astonishing human skills which your memory provides precisely no insight into. You almost certainly have no memory of learning to walk. It's possible you might remember a nasty fall, or more likely being told about having had a nasty fall. But beyond this, learning to walk is – for the majority of us who can – just something we simply can do.

The academic and parenting manuals of child development tell us little about how children learn to walk. Although we read about walking ages, and that a child should hit particular developmental milestones at particular ages (sitting upright, crawling unaided, or attempting to stand upright), we hear little about the transitions between these stages, and why a child who can crawl easily and stably would ever want to make a transition to lurching along in an inherently more unstable position.

The developmental psychologist Karen Adolph and her colleagues have studied in great detail the question of how children make the shift from crawling to skilled walking over a comparatively short period of time. The researchers pose a particularly simple question – 'How do you learn to walk?' – to which they give a graceful and beautiful answer which conveys a world of complexity: you learn to walk by taking thousands of steps and making dozens of falls per day.

Imagine trying to teach a child to walk by verbal instruction: 'Take many thousands of steps per day, and have many falls – oh, and use these falls to learn how not to fall.' It's inconceivable. Or imagine trying to write a computer program to make a robot walk – again, a fiendish problem. It isn't obvious where you should start, and not obvious what the right learning strategy might be. But there are clues. Throughout our evolutionary history there has been a selection bias in favour of those who have been able to learn this peculiar form of locomotion successfully. Those who were clumsy, who slipped, fell, or were caught out by inattention, left fewer descendants. A 'survivor's bias' occurs in favour of those who were better adapted, and walking more efficiently and effectively gradually spreads in the population over time.

Adolph and her colleagues start from the position that quantitatively and qualitatively we know comparatively little about how infants get around; how much time they spend crawling and how much time they spend walking; indeed how they make the transition from crawling to walking; how much they persevere; how intrinsically rewarding and self-reinforcing they find learning to walk. There is little beyond some rough rules of thumb, and to the extent the time courses are understood, they are usually linked to disorders of gait and locomotion, rather than the natural behaviour of walking itself.

There's an even more fundamental question to consider. As a child crawling on your hands and knees, you had a stable posture, one which minimises injury if you fall. From your hands and knees to the ground is not so very far. Why abandon this crawling posture for one that has falls built into it by design? And remember, at first, children are superb crawlers and poor walkers. How do they make the transition from crawling to walking so well and so safely? Walking is unusual as a learned skill. In order to acquire a language children need substantial and sustained exposure to a language environment: in other words, they need to be part of a language community. But feral children, or children that have been abused by being exposed to restricted environments, will still generally walk upright. This suggests that learning to walk is driven more by the unfolding of underlying motor programmes, perhaps inherent in the design of the motor circuits of the spinal cord and the brain, than is the case with language.

In one study, Adolph and her colleagues videotaped the activity of three groups playing in a laboratory under the supervision of their carer.[1] They studied expert twelve-month-old crawlers who had not made the transition to walking; twelve-to-fourteen-month-old novice walkers; and nineteen-month-old skilled walkers. They used a complex laboratory playroom which had a pressure-sensitive gait-tracking carpet for measuring walking and crawling, as well as a whole variety of toys, a stepped ramp, and other stimuli. This experimental strategy generated vast quantities of visual data records of the behaviour of the infants as they crawled, walked, fell, stumbled and explored.

There were several discoveries. Infants vary in their expertise at crawling and walking: some are simply better than others. Infants who are more proficient at crawling covered more ground in less time. Their overall level of activity per se is not important for walking or crawling; more important is the amount of ground covered and how efficiently the infant covers it. Those with greater levels of expertise at crawling are also better walkers – they fall less, stabilise more quickly when standing, and can walk greater distances. This is probably because proficient crawling aids the development of widespread muscle groups that in turn are important for walking. Crawling on all fours involves sequential movements of the arms and legs, and these movements must be coordinated across the expanse of the body trunk. Placing weight on the hands and knees provides a form of ongoing resistance-training for the hips and shoulders, and very likely contributes to the rotational movements that the legs and especially the arms are capable of engaging in during later life.

More than this, though, the coordinated movement of the legs and arms requires muscles to be activated in very precise sequences and groups (technically known as 'muscle synergies') driven by a command signal from the brain; in turn, the feedback from the movement helps sculpt the precision of the movement itself.[2] It also strengthens the muscles of the hips, shoulders, and back which become weight-bearing, allowing the brain and spinal cord to engage in the very precise limb, trunk and head positioning required during the course of walking. Falling was common, with one infant recorded as having fallen sixty-nine times over the space of a single hour. Overall there was no difference between crawlers or walkers in their locomotor skill, when they walked or crawled along a straight, uniform path.

What also turns out to be true is the nearly century-old finding that walking skill (measured as step length and step width) improves with age. Walking is a behaviour with a particular function: it allows you to get from one place to another under your own steam, at your own speed. This is reflected in the finding that 'functional skill' was central to activity levels. Infants who took more steps and who travelled further, fell less. In other words, how well you learn to walk – the complex task of integrating inputs from the senses and the outputs of the limbs ('sensorimotor' integration) that we all take for granted – predicts how often you will fall and how far you can walk. The better you are at walking without falling, the greater the distance you can actually walk, and the less likely it is that you will fall. In other words, in walking, as in so much else in life, effortful and attentive practice makes perfect.

On average, a toddler takes 2,368 steps, travels 701 metres and falls seventeen times per hour. Walking, like the learning of so many other skills, is best learned using a schedule of variable and intermittent practice. Over the course of 300 or 400 days, infants might take many hundreds of thousands of steps and have perhaps many thousands of falls, with the number of falls decreasing in frequency with age and practice and feedback from failure. Infants and children also sleep a lot and sleep is required to solidify the learning that occurs during the day.

Buried in these numbers too is the reason why computationally intensive robotic approaches to learning to walk have been such a failure for so long. The extended period that we have as children – playing in complex terrains with slopes, hard surfaces, soft surfaces, objects of variable dimensions in terms of weight, texture and transportability – provides the training opportunities needed to walk with completely unconscious adult levels of skill.

But we are not just minds immobile in the silent vat of our skulls: we are minds in movement, and we find movement intrinsically rewarding and motivating. So, the developmental move from crawling to walking illustrates in a deep way the theme of cognitive mobility as necessary for us to fully understand and participate in our physical and social worlds. The experience of walking, of movement, is the experience of a brain and mind moving through the world. And this movement in turn changes our experience of the world because the mechanisms of brain and mind are more fully engaged by movement.

The child, grasping at the furniture to pull itself upright, immediately experiences a new world. It can grab things that were previously inaccessible. The child will learn to climb trees, walk upstairs, amble about the garden, wander to school. The child's head will be aloft, and in turn, the child will see the world in a new, and cognitively mobile, fashion. No longer will things be distant; they will walk to them, grasp them, inspect them, perhaps carry them away and break them. Get your collection of fine china and hot saucepans out of reach. These are now things the child can walk, grab, break, spill. And early walking does bring its dangers. Children need to learn where they can walk safely, and where they cannot. And this, of course, is where adults can provide explicit training and teaching on how this wonderful new skill can get you into trouble, but just as easily you can now walk away from trouble.

*

Once we've learned to walk, we walk upright, heads up, leaving our hands and minds free, allowing us to scan the far horizons. To remain upright requires balance: we walk across the earth with coherent motion and direction, keeping ourselves perpendicular to the walking surface, adjusting to changes in surfaces and traction. When on ice, we adapt, we slide and skate, and our balance and heading system continue their work. Why don't we fall over? How do we put one leg so reliably and rhythmically in front of the other? The unexpectedly complicated science of *how* we walk upright is only now being revealed.

The key is in that word 'rhythmically'. There is a rhythm to walking, one we are hardly aware of, unless we pay close attention to it. Our legs reliably change position, one forward, the other stiffening, then the other forward, and the one stiffening, and all the while moving us on, while we remain upright. We'll return to the question of rhythm later. For now, let's agree that walking is an astounding neuro-musculo-skeletal achievement, one requiring the rapid coordination of the activity in your brain and nerves to successive sequences of contracting and relaxing muscles and muscle groups. There are at least ten of these in each leg, attached to your rigid skeleton via tendons. And don't forget the heart is also a muscle; walking makes demands on it too.

How does your brain control how you walk? To do so, it must achieve at least two separate things: balance you in an upright position, and then enable you to move across the ground. But the difference between thinking and doing is more complex and much less distinct than you might suppose. Recall our odd friend the sea squirt, which consumes its own proto-brain when it no longer requires it and becomes sessile. One interpretation, therefore, is that we need a brain for movement, a brain that quickly, automatically predicts – even imagines – the possible movements we might perform. This implies that the brain circuits used for imagined walking – where you imagine walking along a particular path, with particular footfalls – are more or less the same as the ones that are used for actual walking. One important study found that a core locomotion network is active in the human brain during both real and imagined walking.[3] What changes, when actually walking, is that several additional brain regions concerned with motor movements become active.

This offers an important insight into how we imagine things: acts of imagination are exactly that: *acts*. They are neural acts with activations in the brain that can be detected. Imagine a rose: increases in activity in the visual areas of the brain are required to support the picture of the rose you have in your 'mind's eye'. Imagine picking the rose and smelling it: now you have activity in the motor areas of the brain and in the areas concerned with smell. And now imagine pointing at a bird in flight: again, activity occurs in all of the motor brain areas involved in pointing, as well the visual areas of the brain involved in imaging the bird. And imagine your regular walk to work, or in the park, or to the shops. Activity will occur in the motor areas of the brain involved in planning the steps you will make – but there is a twist: there will also be activity in the parts of the brain (the extended hippocampal formation) that support memory, imagination and even mental time travel.

Some things are absent, of course: the actual behaviour itself (in this case, looking at the rose, or the bird, for they are not there; or walking, for you are imagining the landscape you are traversing); what's missing, too, is activity in the brain regions responsible for the execution and control of behaviour (picking up the rose; pointing at the bird; walking to the shops). It is an open question as to whether activity in those brain regions is actively inhibited or suppressed, or if the brain, during imagination, merely does not entrain those regions fully. Perhaps both occur. The deeper point, however, is that imagining things, imagining doing something, and real-world actions are supported by activity in the same brain regions.

It is, of course, one thing to imagine walking. What about the actual 'doing'? As far as your brain is concerned, your body hangs down from your head, until it makes contact with the ground through your feet. You're not built from the soles of your feet up – it's more like your head is a 'castle in the air', with scaffolding reaching down to the ground. Standing upright requires control of body posture: signals from the brain via the spinal cord are required to keep the appropriate muscle control, ensuring you don't topple over. This is a continually monitored process, because your line of gravity is typically slightly in front of your ankles and knees: imagine a plumb line hanging from your chin to your feet. Your line of gravity shifts forwards and backwards in concert with your movements as you walk.

So, from the brain's point of view, the first imperative is to keep body and brain stable in space, whether standing still or striding along. If you fell over, or felt like you were falling over every time you stood up, you will not get very far, and will most likely injure yourself. The head acts as an 'inertial guidance platform' – a platform stabilised for movement, so that it maintains a relatively flat and parallel-to-ground position despite changes in the terrain. They are widely used in aircraft, in submarines, in cars: in a similar way, your brain could just as easily be using a wheelchair or bicycle to get about. But whatever the means of transport, the brain must stabilise the head in space during movement, or walking, cycling and driving will all become impossible.

How does this job of balancing while moving – inertial guidance – work? Draw a line from the corner, or outer canthus, of the eye – the point at which the eyelids meet – to the ear canal. However active we are being, the brain will always attempt to maintain this imaginary line approximately parallel with the ground.[4] This stability of head positioning is achieved by a complex mechanism using inputs arising from the act of moving itself, and the messages coming back from the parts of the body in motion. Surprisingly, it does not use or need an external reference input, such as sight or hearing. This does away with the need for continually monitoring and interpreting the outside world: you don't have to judge every step you are about to take by watching the ground closely. Instead, speed, direction of movement and feedback from the body itself provide all the signals needed. This is why the visually impaired, for example, can still walk with great ease.

Sight and hearing are then used to calibrate or correct errors made by the internal system. In a (limited) way, walking is similar to driving a car on cruise control. The car maintains its constant speed not by the visual world flowing quickly by, but by ensuring that rotational force (torque) turns the wheels at a constant rate. The movement signal originates from within the car: from the depression of the accelerator pedal controlling the drive train of the engine. Movement is not controlled by any external input. And, as we all know, these inputs can be ambiguous for providing speed, depth and movement information: in white-out conditions humans are notoriously poor at telling which way is up and which way down, how near or far things are, or where sounds are coming from. Closing your eyes and feeling the inputs from your body can help reduce the ambiguity.

In humans and all other animals with a spinal cord, the mechanism to stabilise the position of the head and movement is based in the inner ear – it is known as the 'vestibular system', the word vestibule meaning an 'antechamber, hall, or lobby next to the outer door of a building'.[5] If you were foolish enough to push a pen about one to two centimetres through your eardrum, you would permanently damage it. Resist the urge to do so.

But the vestibular system can be disrupted temporarily by spinning in circles – or by consuming too much alcohol. The spinning feeling when a drunk person lies down can usually be relieved by placing a foot on the floor. Contact with the ground is an external input that stabilises the vestibular system – and thus stops your head spinning – by providing the necessary 'proprioceptive' signals (from the relative position of joints involved in movement: hip, knee and ankle). The same is true of a car – hoisting it off the ground and gunning the engine will spin the wheels, but you won't move anywhere. Contact with the road provides the resistance needed for forward motion.

The mechanism of the vestibular system is a miracle of micro-engineering. It has two principal divisions: the semicircular canals, and the otoliths, through both of which fluid circulates. Within the semicircular canals, little hairs protrude from the inner surface. They have small crystals at their tips and are attached to 'stretch receptors' at their base. The hairs move in the fluid. Think of them like tulip flowers at the end of a stalk, swaying in the wind. The minute tugs when they move change the shape of these receptors a little – just as the wind blowing the tulips will cause the roots to stretch in the soil. This stretching, in turn, causes a change in the receptors' electrical state, and sends a signal to the brain via the vestibular nerve. It is a simple, reliable and robust way to convert a movement signal into an electrical signal.

The otoliths, meanwhile, are fixed at right angles to each other, and are hive-like masses with hairs that are fixed in calcium crystals. These align with linear movement: forwards/backwards and sideways (left/right). The crystalline masses move when the head moves – think of a tube containing ball bearings being shaken forwards and backwards or from side to side. And because the position of these always-on senses is fixed within the head, the brain has a constant reference signal provided by the semicircular canals and the otoliths for the brain and body in three dimensions: up/down, side to side, forwards/backwards, as well as rotations about these dimensions.

The vestibular system is locked away deep inside the skull, processing information about what must be happening in the outside world, without actually interacting directly with it. The vestibular system is a fixed sensory system, locked to the position of the head. Imagine a tube containing ball bearings is glued to your hand: the ball bearings move with your every hand movement, just like the vestibular system. Imagine now a slightly more elaborate set of tubes glued to your hand: one tube that parallels your fingers, another across the palm of your hand, and a final one that encircles your knuckles. As you move your hand in any direction, the ball bearings will clack about, faithfully reflecting the speed and direction of your hand movements, without ever touching your hands themselves. Hook these tubes up to measure the movement of the ball bearings, and in principle, you've created something that functions very like the vestibular system: the movement of the ball bearings can tell you about the motion of the hand, without them actually touching the hand itself.

Although the vestibular system provides us with direct, non-visual feedback on our movements, it doesn't always work perfectly. It is well known that when climbers are caught in avalanches, they have great difficulty in telling up from down. This is because they have no reliable external inputs, and the self-generated ones may be rendered insufficient because you might be trapped or in pain. Sometimes climbers will dig in the wrong direction, convinced it is the right one. The trick to know which way is down, should you need it, is to drool some saliva out of your mouth. If it runs into your nose, you are upside down. If it runs down your chin, then you are the right side up – irrespective of whether or not you actually feel that you're in the correct, upwards-facing, feet-at-the-bottom orientation. Because gravity doesn't care: spit is going to run down towards the centre of the earth, no matter what. The same rule applies if it runs sideways across your cheek: dig in the opposite direction.

As the core of our sense of balance, the vestibular system ensures the stability of the body over differing surfaces – a considerable achievement for the brain. It is a silent sense, but one that is always active, even when we are at rest or asleep.[6] There are many ways we know this. Our near relatives, monkeys, are perfectly capable of sleeping in trees, for example; occasionally, they are observed to slip or move, and they right themselves quickly, hopefully grabbing a branch in order not to fall.[7]

The fastest way to wake somebody up, of course, even in the deepest of sleeps, is to cruelly push them out of bed. They will flail and grab the air, or the bedclothes. The persistent activity present in the vestibutar system is known as 'tonic activation', and much like the connection of your house to the electrical grid, it is always on. You only notice the loss of electricity during a power outage, or the loss of the vestibular system if it is damaged in some way. What this activity provides is a consistent input to lots of different brain regions, especially ones concerned with movement and arousal. When the vestibular system is sharply and suddenly activated during sleep, you wake up. But you will generally only wake from movements imposed on the body: your own kicking and moving during sleep tends not to wake you (the movements of your sleeping partner are a different matter entirely, however). The huge evolutionary value of waking quickly during earthquakes, or being able to sleep in trees to avoid land-based predators, is obvious. The key point, though, is the speed and rapidity with which the vestibular system acts, faster than we are conscious of.

The speed of action of the vestibular system is also apparent during walking: let's imagine you go out for a walk on an icy day. You're striding along – and you hit a patch of ice, and immediately try to stabilise yourself. Having managed not to fall, you replay the sequence of events in your mind. You were striding comfortably, then your foot hit ice and you slid. The first thing you will have done is stiffened: all of the muscles of your legs and then your body trunk rapidly acted in concert, and tried to stop your body from slipping or moving further. Then, having successfully righted yourself, you started to walk forward again, this time attending to the ground. What is really striking is that this rapid stiffening happens absolutely automatically and reflexively.[8] It precedes the conscious awareness that your foot has slid from under you. How does this work – and within milliseconds? The signals from the extremely sensitive vestibular system result in several types of information: 'I'm slipping', followed by 'Stiffen muscles quickly', followed by 'I haven't fallen', followed by 'OK, resume walking – but be careful'.

We sometimes walk in unusual environments that are unstable underfoot, such as while on a boat at sea, or inside an aeroplane in turbulence. Walking while the ground moves – something deeply unnatural and unsettling – can demonstrate how the vestibular system adapts to the inputs it receives. The feeling of motion sickness (kinetosis) arises from the disparity between movement sensed by the visual system, and movement sensed by the vestibular system.[9] While motion sickness is unpleasant, it will usually pass as the visual and vestibular systems come into register with each other. The vestibular system is, therefore, not hard-wired: it is capable of learning because of the plasticity of the nervous system, and in the case of motion sickness, it demonstrates learning and plasticity gone awry. A rare and dramatic variant of motion sickness is the very poorly understood syndrome known as *mal de débarquement* (or MDS), or 'sickness from disembarkation' (from a boat or ship).[10] In MDS, the person experiences a profound sense of movement when they disembark from a boat and attempt to walk on dry land. A feeling of motion often arises for a few moments when stepping off a boat, treadmill or escalator, but in the case of MDS, the feeling of motion is long-lasting, but is often relieved when the body undergoes passive movement, such as being driven in a car. The problem is central – that is, within the brain itself – rather than arising from a problem with the vestibular system at the periphery. It might be the case, for example, that the recalibration that the brain performs as result of motion on a boat is not re-recalibrated when back on stable ground. Land legs become sea legs, and remain so when back on land: hence the relief from feeling sick often felt when the body resumes passive motion again. The vestibular system returns to feeling the world as it was, rather than the world as it is, firmly underfoot. The vestibular system is a miracle of evolutionary engineering, quietly intervening in every aspect of our walking lives, while we (usually) remain completely unaware of it.

*

We have other senses that might contribute to how we walk. Sight is a particularly obvious one. Does the way we see the world, at its most basic – the flow of light across our eyes – affect the mechanics of walking? Our eyes tell the brain when we move forward that we are in relative motion between ourselves and the outside world. This is known as 'optic flow': when we are in motion, on a bicycle, on foot, in a car, on a train, the visual world flows past us as we move through it. Picture yourself walking forwards down a narrow corridor with flanking walls: from your point of view, and more particularly from the point of view of the retina, the walls appear to move backward. Optic flow arises naturally from moving forward. We are usually unaware of it, except at those peculiar transitional moments when we go from walking on steady ground to walking on a moving walkway, for example. At the start and at the end of the moving walkway, there is a sudden change in optic flow, where it speeds up or slows down suddenly relative to our walking speed. That feeling of sway, of disorientation, arises because of the disjunction between the movements we make with our bodies, the movement imposed on the body by the moving pathway and the visual world that we are experiencing. This disjunction can be used to tease out how seeing contributes to moving.

How, then, do we use optic flow to regulate our walking? To understand this requires combining the experience of movement in space along with systematic variations in optic flow. You can get elements of this effect on moving walkways, particularly if they have bright, sidelit advertising hoardings, and intermittent interruptions to smooth movement. But in the lab, combining treadmill walking with large-scale virtual-reality screens provides controlled visual input, either in register with walking speed, or occasionally at variance with it.[11] The question many studies address is how rapidly we change from walking regulated by central pattern generators to visually controlled walking. Sudden changes in optic flow provide a top-down signal that something has changed in the visual environment – a little like warning signs that require immediate action when driving. In other words, there is an important top-down influence on walking generated by visual input from optic flow. It can change how quickly or slowly you walk on a moment-to-moment basis so that you match your walking speed to the apparent visual speed of the environment around you.

Now we know how we keep our balance when walking, and how the flow of the world across the retina helps us: but the question remains – how do we actually move? The answer is rhythm, which is intrinsic to walking. A metronome, used for keeping rhythm, is an inverted pendulum, sweeping from side to side. The classic way of thinking about human bipedal walking is as an 'inverted double pendulum', where the body swings over a stiffened limb during each step.[12] During walking, one foot always remains on the ground, unlike running, where both feet can leave the ground simultaneously. Walking is the outcome of an extraordinary collaboration between top-down control by the brain, bottom-up input from the feet and legs, and a mid-level rhythmical control system based in the spinal cord that functions as a 'central pattern generator' (CPG).[13] Pendulums on a clock are a type of pattern generator – they swing back and forth, reliably and predictably, creating a definable rhythm, which is used to drive the action of the clock. A CPG is a circuit in the nervous system producing regular rhythmic motor patterns; breathing, the slow and steady peristaltic waves that propel food through the gut, and your beating heart are other examples.

CPGs in the spinal cord demonstrate that the brain does not, in fact, control the fine details of all of the activity of the body: headless chickens run around, and headless turtles swim for a while too. Eventually, of course, they stop because of cardiac arrest and blood loss, but the continued movement proves that rhythmic motor movements are controlled by the spinal cord – below the level of the brain. The fact these CPGs exist in the spinal cord may also turn out to be a great boon for treatments for paralysis resulting from damage to the spinal cord – if electronic circuitry can be developed that effectively bypasses the site of damage.

Stable, rhythmic movement is the core of walking, another crucial input is the system that allows foot placement without us paying continuous conscious attention to the foot itself. How does the brain solve this problem of knowing where your foot is, and then placing it against the ground, and levering you on? As we will soon see, the brain possesses an acute sense of extended space, a 'cognitive map' that allows you to navigate the world. But it also has an acute sense of the body that it animates. The brain engages in 'exteroception' (processing information about the outside world from vision and hearing), and 'interoception' (processing information about, for example, hunger, thirst, pain from internal organs). We also have a highly developed sense of the position of our ankle, knee and hip joints in space, and also from signals from the muscles and ligaments: this is our 'sixth sense' – proprioception.[14] Testing your proprioceptive sense is the subject both of many childhood games and neurological tests. One simple test is touching your nose with your fingertip with your eyes closed. The margin of error here should be less than about a centimetre. Another is walking in the dark or with your eyes closed – in an obstacle-free environment this is surprisingly straightforward. Finally, stand on one leg with your eyes open, and repeat with your eyes closed. As you try to maintain an upright posture you become acutely aware of how visual input is integrated with your joint sense. This latter experiment shows how the visual sense and the proprioceptive sense can come together to maintain an upright position.

We usually walk with our eyes open, and proprioceptive information and visual information (optic flow) are effortlessly integrated. This provides our normal walking experience of the visual world. After all, we rarely walk backwards over long distances. When you are walking forward, the feeling of the world slipping behind you is known as 'expansion' flow, whereas when you are walking backward, the feeling of the world slipping in front of you is referred to as 'contraction' flow. The horizon shrinks or contracts as you walk backwards from it, and the horizon expands as you walk towards it. Does our experience of optic flow change as we start to make the transition from crawling to walking and the orientation of the head and eyes changes? One important study by the experimental psychologists Nobu Shirai and Tomoku Imura addressed this question, and demonstrated that children who were better walkers preferred expansion-flow stimuli, and this preference became stronger as the children got older.[15] This makes sense. The idea that children would find expansion flow in some sense more interesting and rewarding is adaptive, where the development of independent walking is concerned.

The researchers also found that infants placed in baby walkers have a different experience of optic flow to children who learn to walk normally. Infants in baby walkers kick downwards with their feet, and generally end up pushing themselves backwards, so their experience while moving in the early months is of contraction flow, rather than expansion flow. Other studies have suggested that children who have extensive experience in baby walkers learn to walk more slowly than children who engage in normal crawling, and then make the crawl-to-walking transition.[16] Beyond this, a broader developmental question is whether learning to walk drives other vital psychological changes, especially how children interact with others. Walking is not a solitary achievement: at its core, it evolved in a social context, involving interactions in small family and other extended groups.

We know that walking confers cognitive mobility, but these findings show that walking also changes qualitatively the social interactions of the child. Being able to move about independently, autonomously, self-directedly, on our own two feet, is as vital for our locomotor development as it is for our cognitive and social development. The transition to walking subtly changes virtually every facet of our psychological functioning. Children learn to navigate their homes, their gardens, their streets, their schools, their playgrounds. They will do much of this exploration on foot, and they will do it without much by way of explicit training or instruction. They learn where they are, and where they want to go, on the fly. How do they learn their worlds? Do children – indeed adults too – have something like a GPS in their brains? Do other animals have something like a GPS too? We will turn next to how the brain maps the world while walking.

# 4.
# HOW TO WALK: WHERE ARE YOU GOING?

Let me take you back a few years: I have recently moved to wonderful enthralling capacious multitudinous polycentric polyphonous London. I'm just getting to know the city. It's before we all had mobile phones. I arrange to meet my friend Ted at Highgate tube station to go for a walk around Highgate Hill, and maybe to explore the cemetery. I take the train from Streatham, and then the Northern Line up to Highgate station. I wait, but he doesn't turn up, and can't contact me. So I decide to walk home. It's a beautiful, sunny Sunday afternoon. I don't have an A–Z with me, so I have to find my way using street signs and dead reckoning. It's about eleven miles, including a crossing of the Thames – about three and three-quarter hours. What are my odds of getting home?

Dead reckoning – estimating where you are likely to be, based on your speed and direction of movement from some fixed and known point – is a process used by mariners and navigators since time immemorial. It is used by ants and homing pigeons and humans, and other species too. In biology, it is known as 'path integration' and, if you keep track of your speed and direction of movement, it allows you both to work your way towards a goal and back to your point of origin.[1] But it is not a perfect process and errors do occur.

Knowing I should proceed downhill helps me find my way. I walk from Highgate to Charing Cross Road. I wander about the second-hand bookshops, head south to Pimlico to cross the river at Vauxhall, and then down through Brixton, toward Streatham Hill and Streatham Common. I'm footsore, but elated. I didn't get lost, although I was walking through parts of London then quite unfamiliar to me. Why? Because I had an approximate sense of where I was heading, I knew roughly the places I should encounter, and because I could establish my position using street signs. And in the dark, without street lights, signs or satnav? I would have had to rely exclusively on dead reckoning, a process which accumulates directional or navigational errors unless it is recalibrated regularly by input from the environment. I might still be wandering about.

Path integration is easily demonstrated by finding your way out of a suddenly darkened room in which you know your position and where some of the obstacles such as chairs are. By making a calculated guess regarding the location of the doorway, you move quickly from a frame of reference centred on yourself ('where I am') to one centred on the environment ('where the door is in the room'). This process relies on the vestibular system keeping us upright, in concert with the proprioceptive system acting as our 'sixth sense', constructing and updating a map of space. It especially relies on memory because you must recall the position of the door – your goal – and avoid the obstacles that might be strewn about the room.

We have described how several ingenious bodily systems combine to allow us to regularly, reliably and rhythmically put one foot in front of the other to get where we plan to go. Let us journey now into the deep recesses of the brain to discover how we find our way, wherever we're going, how we get lost, and how we represent the outside world within our brains.

*

How do we walk anywhere? Common sense tells us that walking involves a combination of visual functions and motor functions. But this overlooks the obvious fact that the blind (even those blind from birth) or visually impaired can walk with purpose and direction. They can navigate a complex three-dimensional space without being able to see that space. They can find their way around complex environments, and can find their way back to a point of origin. And the unimpaired-sighted can do the same when blindfolded.

On my way to work I often encounter visually impaired individuals (usually with canes) travelling by bus or train into the centre of Dublin. I am always in awe of them. How can the visually impaired undertake these complicated journeys with little to no vision, and minimal technical assistance? Lowered crossing points in the footpaths, modified underfoot surfaces, and the sounds at traffic crossings all act as helpful environmental signals. But it takes sophisticated experiments conducted in both the normally sighted and the visually impaired to show us fully how this kind of non-visual way-finding is possible.

Some of these experiments involve walking a blindfolded unimpaired-sighted person along a complex route, and asking them to return to a point of origin, sometimes via a particular path. Complications can be introduced – sounds from different directions, changes in the surface underfoot, rotations of the body to upset the vestibular system. This last manipulation is perhaps the most critical because losing the sense of stability and direction provided by the vestibular system has a devastating effect on our ability to perform a simple, point-of-origin-finding task. We are fooled by our sense of the spatial world as being visual in nature. In fact, as far as the brain is concerned, vision is merely one sense that contributes to our understanding of space – an important one, but just one. And we know this because we can find our way around familiar and unfamiliar environments in the darkness. This sense of space (now known as a 'cognitive map') can be regarded as a 'silent sense': it is constructed largely without our awareness, and we only notice it if it fails us.

Over several decades, a number of significant studies have been conducted on the spatial sense (or cognitive mapping) in patients who are either blind from birth, or who have become blind later in life. These have compared their performance on walking tasks with blindfolded, normally sighted, age-matched controls. They use path-integration tasks to examine whether vision is required for normal, locomotor-based spatial abilities; participants are asked to walk along certain routes or trajectories, and must then either reproduce the route to get back to where they started from, or alternatively find the shortest path back.

If we believe vision is required to develop normal spatial abilities, then we might predict that sighted people will perform better than those who became blind at some point after birth (i.e. the adventitiously blind). In turn, the adventitiously blind should perform better than those who are congenitally blind (i.e. blind from birth). Each group has different experiences of the visual world – from total, in the normally sighted group, to partial and historical in the adventitiously blind group, to none at all in the congenitally blind group. In one particularly enlightening study, a variety of experiments with these differing groups required the participants to produce, reproduce, or estimate short walking trajectories that may involve just a single body turn. In more complicated experiments, the participants walked multi-segmented paths, and had to retrace their routes backwards, produce shortcuts to a point of origin, or point to a place by inferring its location from the known location of some other object.[2]

The striking revelation is that normal vision is not required to learn to walk and navigate in extended three-dimensional space. Overall, performance on the simple tasks was approximately the same across each of the three groups. Whether normally sighted and blindfolded, adventitiously blind or congenitally blind, all participants were capable of covering the short distances to be walked, reproducing them or estimating them to approximately the same level of performance. On the more complex tasks, all groups also performed to a similar level. The congenitally blind were perhaps the worst performers, but there was generally a noteworthy concordance between the performance of the blind and sighted participants.

What is also clear is that having extensive experience of the spatial world is necessary for building usable cognitive maps. Vision may predominate this spatial sense, but our spatial sense is built up from our experience of walking about the world, and is to a large degree independent of any particular sense we use to interact with the world. Our spatial sense is only somewhat like our visual, or our sense of hearing or movement, but it is more abstract and less immediate than any of these because it is substantially constructed from the inputs these other senses provide. It provides a map of the possibilities for movement in the world, giving you the constant 'what and where' that you need as you move about.

What's more, being able to walk in the world, even without seeing, ensures that we are capable of learning about our environment. We know that the hippocampal formation in the brain, within which the cognitive map of the world is inscribed, receives inputs from *all* of our senses, plus feedback from the motor system. While we are usually unaware of how powerful our spatial sense actually is, it is nonetheless there, and it is activated most effectively at walking speed. The spatial sense is a little like an operating system for the other senses: the invisible architecture behind the documents you can see on your computer screen; without it, little works.

But even if we now know that vision is not a necessity for moving accurately around our environment and that vision is only one ingredient that makes up our spatial sense, the big questions remain: how do we know where we are, and how do we know the pathways to where we are going?

The answer lies in work conducted about seventy years ago by the University of California at Berkeley psychologist Edward Chace Tolman, who first introduced the idea of the 'cognitive map', the abstract map of the environment created by the brain which allows us to navigate the three-dimensional world.[3] Tolman studied the behaviour of rats in mazes, and was particularly interested in a phenomenon known as 'latent learning' ('latent' because what the rat has learned is not immediately visible in its behaviour). Tolman chose a simple but revealing strategy. He allowed rats to wander about complex mazes, where occasionally they would encounter a nice morsel of food in a particular location, reached by a particular route, involving particular motor movements – only left turns, for example. He then blocked off certain routes in the maze. What would the rats do on finding their route inaccessible? Would they perform the previous sequence of motor movements that resulted in a reward?[4] At the time, an argument was raging within psychology – one that seems hard to credit now – over whether we (and rats, mice, and monkeys) behave as we do only because rewards or punishments have explicitly shaped our past behaviour. Furthermore, attempts to understand and predict behaviour should be understood in terms of the stimuli that drive responses. This was sometimes known as stimulus-response (or S-R) behaviourism.

Tolman took a different view, influenced by gestalt psychology, an important undercurrent in psychology during the mid-Twentieth-century which focused on how we perceive the world more or less instantaneously as a whole, rather than as components that have to be built up, bit by bit. Although he disliked the English translation, the famous phrase 'the whole is greater than the sum of its parts' derives from the German gestalt psychologist Kurt Koffka.[5] It would be fair to say that Tolman might have wondered if rats were gestaltists. Would frustrated maze-running rats, he speculated, perceive the maze situation as a whole, and base their future actions on what they learned from their previous experience of the overall properties of the maze?

If the behaviourists were correct, the rat would repeatedly attempt to take paths that had previously been rewarded, and be stumped by the new obstruction. If the gestalt view was correct, the rat would quickly figure out an alternative route to the food. And the gestalt view was correct. While they had been freely exploring the mazes, Tolman's rats had incidentally and automatically been learning something about the layout as a whole – they were either learning or inducing or inferring something like a 'survey map' of the maze. Moreover, they were able to use this map to solve problems. Rats and humans, Tolman suggested, possess a 'cognitive map' which is the basis of our understanding of the extended, three-dimensional space we inhabit.

This outwardly simple experiment was a landmark in our understanding of how we navigate the world. It told us that animals (and, presumably, humans) quickly develop internal maps of the world that flexibly guide goal-directed behaviour. Now, where in the brain is that map?

*

At the heart of navigating our world is knowing where we are, where we want to go, and then going there. Without keeping these in register, we will get lost. If we carry a mental map with us wherever we go, what happens when all the signs have gone?

One experiment has addressed this question directly.[6] Participants were asked to walk either in a large and dense forest or in the Sahara desert. Their task was simple: to walk in a straight line for a minimum set period, usually a few hours.

Some walked in the day, others at night. All wore GPS tracking devices. While walking without reliable visual cues in the fog, or with heavy cloud-cover, the subjects regularly veered left or right, and eventually crossed the path they had been on. In clear daylight, they sometimes veered from a straight path but neither systematically walked in circles nor repeatedly crossed their own path. The result was the same in moonlight. In other words, the presence of a large, relatively constant cue in the sky from the sun or the moon allowed them to walk in a relatively constant straight line, with a relatively constant angle between them and the sun or the moon.

In subsequent experiments, blindfolded participants were tested at an airfield. Without a constant external visual input to recalibrate their paths, after the first hundred metres or so people walked in circles of approximately twenty metres in diameter. The lesson here is straightforward: humans have an internal directional sense that works well when walking shortish distances of tens of metres, but over longer distances, without fixed cues to recalibrate our position, we systematically veer off straight courses and often end up walking in circles.

We can investigate how these navigational mistakes happen using virtual reality – a recent boon to experimental psychology and neuroscience. VR allows us to construct complex virtual cities, then ask participants to explore them in order to observe how errors arise. And what gives birth to these errors is the gap between our cognitive map (our 'brain's eye' of the world), and reality: the unforgiving world as it truly is.

To get to where you want to go, two estimates are always necessary: one is a *straight line* or as-the-crow-flies estimate to the goal, and the other is an *on the ground* estimate that may feature extra distances caused by the presence of obstacles.

Even fairly straightforward destinations may require some degree of circumnavigation, where you follow a straight line and then make a left turn or a right turn.

In one recent study, scientists set out to ask whether there are biases in our estimates of the time it will take to get somewhere, or biases in our estimates of the distance it is to somewhere, or some combination of the two.[7] They created a virtual city and asked participants to act as pizza deliverers. Their job was to drive at a constant 35 kph to destinations which they could approach directly, with a minimum of circumnavigation, or which required a considerable degree of circumnavigation. The simple routes were L-shaped, involving a single turn, and the more complex routes were U-shaped, requiring the deliverers to return almost to the point of origin. Overall, participants consistently underestimated the time it would take to get to a goal, and consistently overestimated the as-the-crow-flies distance. What this confirms is the existence of the 'it won't take long to get there' fallacy. We underestimate travel times because of mistakes about distance – and these arise because we underestimate the complexity of the paths that we have to take.

Although the variety of ways in which we can become lost is considerable, we also get navigation right to a surprising degree. What inner capacity is it that allows us to do that? We've all experienced the satisfaction of finding our destination on the fly, despite not knowing the exact route and using only cues from landmarks, signs, and instincts that are hard to articulate precisely, or of returning somewhere we haven't been for years, and discovering that we can still get around. How is this possible? The brain does, in fact, have its own GPS-like system, and here we will explore the many dramatic discoveries that have been made about it.

But the extra factor that helps us find our way is that humans are good at ruminating on our pasts and imagining alternative futures – a capacity that is probably unique to us. The brain's GPS system taps into this and allows us to engage in mental time travel – via memories, or imagining alternative futures. This is a map of time, rather than space, but it is equally essential. People with damage to the brain's GPS system often lose access to their pasts, find it difficult to record ongoing memories, and can't imagine the future. In the elegant phrase of the late Suzanne Corkin, who pioneered the study of amnesia, their lives comprise a 'permanent present tense'.[8]

The sense of time passing is bound up not only with our sense of memory, but also our imagination. Curiously, amnesiacs do not show any great sense of distress regarding the permanent present tense that may seem to the rest of us a kind of purgatory. It is almost as if the loss of their own personal timeline and the loss of imagined alternatives in the future is accompanied by a recalibration of how they experience their world. What this suggests is that the pain experienced through any life-altering event (such as a death of a loved one) might arise in part not from a loss in the present, but also from the loss of imagined futures.

In addition to simulating differing futures and pasts, we are also able to simulate differing routes and paths to a destination. And the key to the mystery of how we engage in our rich, imagined mental lives might partly reside in the activity in brain regions that pre-plan and support navigation in the real world, but which do not directly participate in their execution. The hippocampal formation does not control your leg movements, for example, but it does create the map of the world that allows other brain regions to direct your leg movements that walk you to where you want to go.

To guide us through the bits of brain that we are about to encounter, we need a simple way of thinking about the anatomy of the brain. So, take your right hand, clench it into a fist, and hold it upright, with the elbow bent; now, take your left hand and lay it over it, so that your thumb of your left hand is pointing to the right. You can treat your right-hand wrist and forearm as the spinal cord. Your clenched-up fist is the thalamus, and your left hand is the neocortex. (This is a very rough guide; your brain actually divides neatly into two almost exact mirror images, for example.) Still, your right thumb corresponds to a structure that we will meet time and again, the left hippocampal formation (you have two of them), though ordinarily they would be covered by brain tissue. They run approximately from behind the ear toward your temple. The brain's spatial-cognition system corresponds approximately to parts of the thalamus (where the covered knuckles of your right hand are), the hippocampal formation, and a variety of cortical structures represented by your knuckles and fingers of your left hand.

The functions of these regions of the brain can be probed using a variety of methods. The oldest relies on accidents: happy accidents for the brain scientist, but for the patient, sadly, not so much. Patients may come to medical attention because they have had a stroke, a brain tumour, or an infection which has destroyed part of the brain. They may have suffered an unusual head injury from a fall or from a foreign object penetrating the brain. More rarely still, a whole brain region may have for some reason not grown at all. These are 'agenesis' patients, who can offer important clues as to the functions performed by particular brain regions.

It is rare that a single patient will have damage that is restricted to a particular region which discretely compromises one set of functions. But by carefully compiling patient cases with particular types of injury, some attribution of function to brain regions is possible. Latterly, other methods have become available, including a variety of brain-imaging techniques, as well as techniques to measure the underlying electrical activity of the brain (especially EEG, the electroencephalogram). Combining these techniques with the experimental methods used in psychology has given rise to 'cognitive neuroscience', the attempt to map specific psychological functions onto brain regions and brain networks.

One very famous patient, known by his initials as HM, suffered a road accident as a child which resulted in him becoming severely epileptic. The epileptic focus was also apparently localised to his hippocampal formation on both sides of his brain, which was surgically excised in 1953 in what his surgeon William Scoville referred to as a 'frankly experimental operation'. The result was that HM's epilepsy was relieved but he was left with a grave, non-resolving and enduring amnesia. His story has been told often, so we will not dwell on all of his deficits here. But one deficit is striking. After his operation he became severely topographically agnosic: he became easily lost, and couldn't learn routes through new environments. All our familiar ways of learning about the world were denied to him.

The hippocampal formation is required for learning about where we are in space and, as we will see, walking or voluntary movement activates it, by giving rise to a reliable and repeated electrical rhythm in the brain. In rats this rhythm is referred to as 'theta' rhythm, and its activation through walking is required for spatial learning. Movement is essential to building our knowledge of the world, and the best form of movement for building this knowledge is physical locomotion; walking is probably the best of all because the timescales that walking affords us are the ones we evolved with, and in which information pick-up from the environment most easily occurs.

Theta is an important and interesting biological signature for active movement, at least in the rat hippocampal formation. Theta was long predicted to be present in human brain, because it had been observed in mobile rats,[9] but similar experiments could not be conducted on humans – finding theta in humans remained for special neurosurgical circumstances to arise. Standard methods involve recording the EEG signal present at the scalp in humans, but the location of the hippocampal formation means that detecting hippocampal theta is extremely difficult: it might be sitting in there somewhere, but finding theta, *if it is present*, is a challenge.

The latest generation of miniaturised neurophysiological recording systems, however, allow just this possibility in human patients using electrodes implanted for the purpose of assessing them for epilepsy surgery.[10] What the study shows is that theta is indeed present, and to all intents and purposes, it differs little from theta found in the freely moving and exploring rat. Humans, like all other mammalian species tested, express theta in the hippocampal formation when they are walking about the environment. In other words, theta is something of a biological universal; it is a signal, conserved across species, that the brain is engaged in active exploration and movement within an environment.

*

Walking arises because coalitions of brain areas temporarily organise groups of muscles that allow you to walk. Let's explore how our rats walk about mazes, creating cognitive maps as they go. What is going on in their brains as they do this?

Fascinated by Tolman's important insights into the cognitive map, the neuroscientist John O'Keefe conducted experiments at University College London which started the modern revolution in our understanding of the brain's GPS-like system. He implanted microelectrodes into the hippocampus of rats, and recorded the electrical signals that these brain cells displayed while the rat wandered about simple mazes, looking for food.[11] The electrical signals that single brain cells produce are so small that they have to be amplified 10,000-fold in order to be detected. Usually, the signals are captured on a computer and played through a loudspeaker, while the position of the rat is tracked on a video camera. Some data processing is then performed to produce composite maps of behaviour: these combine the position of the cell firing and the position of the rat on the maze.

Brain cells have particularly distinctive sounds when they are played through a speaker (there are many recordings available online). Some sound like angry bees, going ZZ-ZZZ-ZZZZ-ZZ-ZZ. Others sound like a dying wasp, making a zzzZZZZZZZZZzzz sound, then falling silent before suddenly coming back to life. Listening in to the ongoing neuronal chatter is a sobering and amazing experience. You know you are listening in to a brain cell close to the electrode tip, and it is in conversation with its neighbours near and far: you are experiencing something – hearing something – normally inaccessible from within the deep, dark silence of the brain. Decoding these sounds and signals is the path to understanding what the brain does, cell by cell, neuron by neuron.

O'Keefe made an astonishing, Nobel Prize-winning discovery: single cells in the hippocampus were mostly silent, except when the rat explored a particular region of the maze. These cells fired in a particular *place* – they are interested in *where you are*, not in what you are doing. The rat would walk into a particular part of the arena, and a silent cell would blaze into life. If it remained in that location the cell would continually fire away, broadcasting the rat's location to its cellular friends and neighbours. Recording the activity of many of these hippocampal cells by using several electrodes at once shows a 'Whac-A-Mole' phenomenon: while the rat walks about, one cell falls silent, another comes online, and so on. The complete picture of hippocampal activity covers the whole of the environment in question. And walking is key (rats pulled around on little trolleys show dramatically reduced activity in the hippocampus).

Using microelectrodes implanted in the human brain, we have learned that we have place cells too.[12] They have come to be recognised as the core elements of the cognitive map – they tell you where you are in the world, and they work best, and acquire most information, when we are walking.

Place cells can also sometimes code for which way the rat is pointing. In rats, place cells are often recorded on 'radial mazes' – an experimental set-up with a central hub, the rat's task being to walk to the end of the corridors of the mazes and retrieve food pellets. Under these circumstances, place cells have directional preferences: they will fire in one direction but not in the other. Part of the problem when we are lost, whether hillwalking or while in an unfamiliar part of town, is that you are likely to have been walking in a single direction. Your place cells will fire when you are walking in one direction but not the other. And when you realise that you're lost, it's because you aren't getting the constant feed that hippocampal place cells provide.

The activity of these place cells can also be seen during brain-imaging experiments, with a huge level of activity occurring in the hippocampus during the exploration of virtual mazes.[13] Many early experiments in imaging the brain's activity during navigation and exploration relied on an unexpected source: the development of visually rich, three-dimensional 'shoot 'em up' video games. Games such as *Soldier of Fortune* or *Doom* involve extended navigation across difficult terrain, with often deliberately poor lighting, dead ends and other challenges. Serendipitously, they also provide an almost perfect window into the navigation and latent learning needed to solve complex mazes. The added realism of occasionally getting killed might even recall stalking wild beasts all those millennia ago ... In these experiments, the subject lies in an fMRI scanner, and plays a stripped-down version of the game. Activity in different brain regions is measured in turn, and a signal is reliably found around a network of connected regions centred on the hippocampus.

O'Keefe's initial description of place cells must have seemed at the time almost miraculous. Here were cells deep in the brain whose activity could be recorded, listened to, and observed in real time. When O'Keefe conducted his experiments, few labs in the world were capable of recording the activity of single brain cells in the behaving animal, or indeed in humans. Usually, these experiments had to be conducted in deeply anaesthetised animals. For some years O'Keefe's work was perhaps regarded by some with benign indifference and incuriosity. Slowly, though, the evidence he produced led other scientists to follow in his footsteps.

James B. Ranck Jr., in New York, was one such. Ranck had also conducted recordings in the hippocampal formation but had missed the existence of place cells because he recorded in small and restricted box-like environments, which offered the rat comparatively little room to walk and explore.[14] Ranck, however, was subsequently to make an astonishing discovery of his own when he described what came to be called 'head-direction' cells.[15] These are cells in another part of the brain, the dorsal presubiculum, adjacent to the hippocampal formation, but not anatomically part of it. Head-direction cells signalled directional information about the rat's head position, rather like a compass. Head-direction cells are also behaviour-independent, just like place cells, and they fire depending on the animal's orientation, not on the basis of rotational or translational movements of the head. They are not interested in what you are doing – they are concerned with where you are oriented in the environment.

Here at last we are now starting to assemble the elements of a proper cognitive map in the brain. Place cells code for where you are in your environment; and head-direction cells code for your orientation in that environment. In other words, there are two populations of brain cells that are directly involved in knowing where you are and where you are going. Beyond that, since the early 2000s a staggering range of cells throughout what might be referred to as the extended hippocampal formation system have been described. In the entorhinal cortex, for example, the Norwegian researchers Edvard Moser and May Britt Moser discovered cells which have become known as 'grid' cells. These answer the long-standing question of how the brain knows the dimensions of space at all: at least over short distances, entorhinal cortex grid cells appear to provide a metric that the brain uses to code for distance. O'Keefe and Moser and Moser shared a Nobel Prize for their work on the brain's GPS system.[16]

It is now known that the brain's GPS system is distributed across multiple interconnected brain regions. Moreover, there are many other cell types contributing to our sense of space. Place cells signal where you are; head-direction cells signal where you are heading; the boundary cells of the subiculum, anterior thalamus and claustrum tell you about the edges of the environment; and grid cells provide a metric for space. In addition to the core GPS system, there are other cells that signal distance to objects, speed of head and body movement, and your relative upright position. So now our picture of the cognitive map is becoming more complex. In recent years, many other cell types have been described, and one review suggests that there are head-direction cells in at least nine separate brain regions.[17] Additionally, the brain also has multiple representations of the perimeters of environments, cells which fire only when the animal is adjacent to an impassable boundary: vertical walls, or vertical drops which you cannot walk over.[18]

In my own research, I have focused on understanding the cells that signal where you are heading while moving, the position of boundaries in maze environments, and the presence of immovable objects in the environment. We have discovered perimeter cells (which signal boundaries) in two distinct brain regions. One is the rostral thalamus, which connects extensively with the hippocampal formation; the other is the claustrum, a mysterious and thin sheet of cells oriented toward the front of the brain.[19] Cells have also been found which respond to the presence of objects within the environment. These combine sources of information from sight, touch and place to code for particular objects: they are 'multisensory' cells. Other cells respond to goals that the rat might be navigating toward: they fire as the rat approaches a food reward in a maze.

The GPS network in the brain makes possible the coherent and directed motion of the body that we recognise as walking. The brain's GPS system is also found in widely divergent species – it is 'conserved' by evolution. Beyond extending the boundaries of what we know about our brains, these discoveries have also led to the development of 'biologically plausible' robots, which attempt to solve navigation problems in similar ways to biological brains. The brain's navigational and mapping and memory systems are so intertwined as to be almost one and the same. Walking to somewhere depends on the brain's navigation system, and in turn walking provides a vast amount of ongoing information to the brain's mapping and navigation systems. These are mutually enriching and reinforcing systems.

What we cannot divine yet is whether there is a whole panoply of other cells waiting to be described by neural cartographers. Head-direction cells are found in at least nine brain regions. Place cells, by contrast, have only been described in three, and grid cells have only been found in two. We have no good theory to explain why the brain should maintain these multiple codings of head direction in so many diverse locations. My own view is that the importance of the head-direction signal has been underestimated within brain science and that head direction plays a subtle and not yet well-understood role in ongoing cognition. If you observe rats playing, foraging or scrambling around an environment, one of the things you notice is that their heads move about more or less continually. They are engaged in what might be referred to as continuous information pick-up about what is going on around them.

What of the human head? We have a very mobile head, and our eyes move independently within it. Spend a little time watching humans interacting and it is apparent how our heads are continually undergoing subtle changes of orientation, while the eyes also move constantly, allowing continuous exploration of the environment. Humans and non-human primates have an especially elaborate system controlling the movement of the eyes in the head, and also for deploying attention to interesting things in their environment, independent of eye position. From this point of view, therefore, the multiplicity of signals that code for head direction is not such a surprise. The head direction is probably present in so many brain regions in order to allow the rapid movement first of the eyes, and then the head itself, away from or toward things of interest, and to allow quick decisions to be taken.

O'Keefe's work, and that of the experimenters who followed, reveals a profound and intellectually satisfying understanding of how the brain codes for three-dimensional position in space. Firstly, the brain *does* indeed have a GPS-like system. The brain's GPS codes for your position in the world, independent of what you are doing in that space, and allows animals and humans to solve problems that are absolutely fundamental to survival: to find and remember secure and safe places for shelter, or to find reliable sources of food, for example. This system is activated by movement, such as walking or running. It has also been co-opted in the case of humans to allow mental time travel, in addition to supporting physical space travel. It even allows one to deal with the problems presented by predators in the environment. You learn where possible refuges are, and you know where the boundaries of those refuges are, because of 'latent learning' during exploration; you also learn to deploy that information quickly and effectively in order not to become a meal.

Our walking journey so far has taken us through how we find our way, how we get lost, and how we build our inner maps of the world we inhabit. The world we inhabit is increasingly urban, and it is a complicated one. Our urban space is a world of artifice, quite unlike anything found in the natural world we evolved in. Most humans now live, and therefore walk, in a built environment – their village, their town, and particularly their city. Let us now walk the streets of the city, and see how it moulds our walking.

# 5.
# WALKING THE CITY

Walking a city is the best way to get to know it. You can't get to know the mood of a place, its energy and pace, when you're driving or being driven around. On foot you are directly in touch with city life in all its dirt and glory: the smells, the sights, the thrum of footsteps on pavements, shoulders jostling for position and placement, the street lights, the snatches of conversation.

I'm not alone in loving a good city. Perhaps the most famous city dweller in literature is Charles Baudelaire's *flâneur* – a casual wanderer, an observer and reporter of nineteenth-century Paris. Our cities have, of course, changed a great deal since then – for one thing they are now dominated by traffic, leaving the *flâneur* waiting for the green signal at pedestrian crossings. But even more dramatically, more than half the world's population now live in urban areas, and projected estimates indicate this trend will only grow. The latest UN projections expect the world's population to grow by 2.9 billion in the next three decades, and possibly by a further 3 billion by the end of the century. By 2050, it's likely we will be a largely urban species, with 80–90% of people living in towns and cities.

So what does this dramatic urbanisation mean for our ability to walk around a city? How easy is it to be on foot in an urban environment? What is the real experience of city walking? Walking in our towns and cities of course presents a different set of challenges to walking in the countryside. When you hillwalk, you often follow paths that have been inscribed by millions of footfalls over very many generations. That is not true of our modern towns and cities. Where we walk, and the surfaces we walk on, are specified not by prior natural human movement, but by deliberate design and engineering. The pavements vary in their texture and design. Some are poured and moulded concrete; others, large slabs. Somebody has to sit down and think about them, design them, and place them in such a way that they become, hopefully, a beautiful part of our walking environment. Should they be slip-proofed for frosty weather? Will the path offer some spring to facilitate walking, or resistance? And, of course, they have to be paid for, which means taxes must be collected and spent appropriately.

In far too many cities, though, managing vehicle flow is the priority of the urban planner, and ensuring that our cities are walkable is something of an afterthought. It is as if engineers and others conceive of our lives as being contained in boxes: moving boxes (cars) and static boxes (buildings). Walkability is reduced to the short transition zones between these boxes. And in some respects they are correct. We spend most of our lives in cars, buses, trains or buildings, and relatively little time with the air and natural light on our faces. This lack of exposure to nature is something that arises naturally because of the design of our built environment, unless we take deliberate steps to countermand it.

Walkers are corralled by planners at crossing points which restrict natural movement, and this is why walkers often end up forging their own paths contrary to engineered paths. These trails and shortcuts are 'desire paths' in the words of Andrew Furman, the designer and architect, while the writer Robert Macfarlane calls them 'free-will ways'. Macfarlane writes that these are the 'paths and tracks made over time by the wishes and feet of walkers, especially those paths that run contrary to design or planning'.[1]

We are still learning the lessons of urbanisation, and how it affects every aspect of our lives. And yet urban design is something owned and practised by architects and city planners rather than by neuroscientists or psychologists. This is a great pity, something to be lamented, because the science and sensibility that psychology and neuroscience can bring to urban design – to improve the liveability and walkability of a city – is significant, as we will see. Urban design that fully and properly takes account of the needs of walkers will make cities much more attractive places to live and work. Churchill famously said that 'We shape our buildings, and then our buildings shape us.'[2] Similarly, we first shape our cities, and then our cities shape us. To extend the metaphor, our cities walk us, for the shape of the cities we create determines the shape of our urban walking, for better or worse.

There are great potential walking futures in cities for us all. What we need are acts of imagination fusing the needs of walkers with the expertise of town planners and that of psychologists and neuroscientists. In turn, science, imagination and evidence need to be turned into policy and from that point translated to beautiful, interesting streets of ease, variety and quality. Road-crossing designs, street furniture, the texture and type of footpaths and pavements, the presence of cars and buses – these all act for or against our ability to walk in cities.

Some cities have a porous and fluid quality that makes walking around them a joy. Some cities are awkward, uncomfortable, exhausting, even dangerous to the walker. A useful tool for thinking about how we can walk around in our environments, and in particular how we can compare one environment to another, is to create a 'walkability index'.[3] This can be devised in a number of ways, but a simple way of thinking about it is that it should measure how easily you can do all of the jobs of everyday life on foot, compared to other types of transport.

A very walkable city is one where, when you go out of your front door or hotel lobby, amenities are all within at most a few minutes' walk. Weather permitting, you can stroll to and from your local restaurant or local school. High walkability allows you to engage in the activities of daily life as much as is possible on two feet, rather than having to resort to driving. Some cities, or some parts of some cities, make this much more doable than other cities. Bologna in Italy – a wonderfully walkable city – was commended by the famous writer Umberto Eco as being 'all texture and no excrescence ... it is a city of communal spaces, arcades, bars, shops, a city whose sight lines are designed to meet shopfronts, café tables and other people's eyes'.[4]

At the other end of the scale are car-dependent cities, where just about all activities require a car. In a study conducted on a selection of cities, it was shown that the higher the walkability of the city, the lower the activity inequality (a measure of the degree to which each person walks a similar amount as other people; it is a similar measure to income inequality – the extent to which incomes are the same or different in a population), meaning that overall population obesity was also lower.[5] A comparison within a single US state drives this point home. Three cities in California (San Francisco, San Jose and Fremont) have a similar climate, not dissimilar levels of affluence, and relatively similar demographics. San Francisco scores as a much more walkable city than either San Jose or Fremont, and activity inequality in San Francisco is the lowest. It is no surprise (when considering the USA) that New York, Boston, and San Francisco are among the most walkable of US cities, and correspondingly have the highest levels of walking and lower levels of obesity, on average. The reverse is true, of course, in cities which are of low walkability.

'What works best in the best cities is walkability,' says Jeff Speck, the renowned urban planner.[6] And the best walks in cities, according to Speck, must be useful, safe, comfortable and interesting. For a walk to be useful, according to Speck, 'most aspects of daily life are located close at hand and organised in a way that walking serves them well'. Walking should be safe: this is self-evident, although sometimes ignored. Pedestrians should not be put at risk by fast-moving vehicles and should be treated by urban engineers with at least the same respect as traffic. (Imagine if we invested as much in walking in our cities as we do in driving in them!) Walks should also be comfortable, and here Speck outlines a powerful idea – that city planners and designers should think about urban streets as being akin to 'outdoor living rooms'. We, the users and walkers of cities, should find the streetscape welcoming, with entertainment, seating, refreshment and diversion available. Finally, walks should be – must be – interesting. Speck suggests that to be interesting, our streets should have 'unique buildings with friendly faces, and that signs of humanity abound'.

The most obvious and often the most enjoyable space to walk in the city, and one that usually ticks all four of these boxes, is green space. Is there any better urban space to enjoy than London's Hyde Park, Dublin's Phoenix Park, New York's Central Park, Paris's Jardin du Luxembourg, Cubbon Park in Bangalore, or the Villa Borghese Gardens in Rome, to name just a few? However, with urbanisation set on a one-way trajectory, a common and reasonable fear is that urban development has and will encroach upon historic areas of urban green space, meaning that trees and hedging that reduce the urban heat island effect are being lost. Western Europe has had an urban culture of great parks which has developed over several centuries, and which has sustained itself through famine, disease, and warfare. Are these green spaces under threat from a tide of concrete and tar?

In a study that examined green spaces in 366 European cities with a total population of 171 million people, researchers found a dramatic variation in the amount of green space available.[7] In Reggio di Calabria, in Italy, green space comprises approximately 2% of the urban space, whereas in Ferrol, in Spain, it is 46% of the available space. Is this because some cities build over their green space when they expand? Or does green space have a 'sticky' quality that means once a green space becomes green it tends to stay green? According to the study, what appears to happen is that the intensity of building use rises (building on existing space), rather than additional buildings being constructed (which would expand the amount of built-on space), meaning that green space actually remains relatively stable despite changes in population. This is good news, in the sense that easy access to nature is a vital contributor to, and supporter of, our mental health. However, as the urban population grows, the green space available per person necessarily falls.

Besides the mass move to cities, perhaps the other most important demographic change is that populations are ageing. In general, people are living longer than at any point in history – worldwide, the average life expectancy is now 71.5 years, compared to a life expectancy of a brief twenty-five years during the Roman Empire.[8] Most of this ageing population will want to, or will have to, live in cities. But will it be *safe* for them to live in our cities? Our bodies and brains change as we get older, including a slowing of walking speed and reaction times. Taking account of the inevitable frailty that comes with human ageing should be an imperative for the design of our towns and cities. The newly elderly may find themselves unable to cross roads quickly enough, or find they are unable to get to shops; they may require mobility walkers and other assistive technologies. Simple design problems may also hinder them, as for example, when pavements aren't lowered at crossing points to allow them to cross roads and streets easily. We will find we have condemned a generation, and the generations following them, to being trapped in their own homes, with the loss of autonomy, personal dignity and individual and social well-being that this brings.

An important study in the UK of 3,145 adults aged sixty-five and over tested for impairments in walking speed to investigate if simple tasks like crossing streets at signal-controlled junctions might be difficult or impossible as a result of ageing.[9] It was found that 84% of males and 94% of females tested had a walking impairment. Road crossings are usually set for people who can walk at least 1.2 metres per second. Yet the vast majority of the older adults tested walked at below this speed.

This means they may be only able to cross roads safely if traffic volumes are sufficiently low. Unsurprisingly, older people, as a proportion of their age demographic, are more likely to suffer road-traffic accidents.

The number of people aged sixty or older is expected to increase from one in ten to one in five by 2050 (that is, 2 billion people).[10] Our ageing society presents a challenge for urban design, without doubt, but small, marginal and continual adaptations in walkability design can pay huge dividends both for society at large and for individuals in particular. Changes that make walking life easier for the elderly and disabled also make walking life easier for us all. Walking ramps and lowered footpaths equally assist those on crutches, those in wheelchairs or other mobility devices, as well as children being pushed in prams or strollers. A more walkable city, in straight, is a city that benefits us all in so many obvious and occult ways – obvious, because walkability adds to our health and well-being; occult, because walkability has so many hidden benefits for creativity, productivity and enriching our societies.

*

Making our cities walkable isn't useful only for ease of movement without a car; it has many positive, though not immediately obvious, spillover effects on our societies at large. Densely populated but walkable cities minimising urban sprawl are more economically and environmentally sustainable. They reduce transport costs and transport time, and collaterally ensure that city dwellers walk around. In turn, this movement itself has the wonderful benefit of maintaining and even augmenting brain and heart health, and, as we saw earlier, lowering chances of obesity. This can also correlate to an overall increased well-being in urban residents, reductions in crime and greater social cohesion.

We have come a long way in the development of our urban environments. The streets of our cities and towns a century and a half or so ago presented a considerable public health risk. Open sewers were common, indoor toilets were few and far between, and human waste and effluent handling were of an especially poor standard. Observing acts of necessity that would appal us now, John Gay's 1716 poem, 'Trivia, or The Art of Walking the Streets of London' warned of the dangers posed by the emptying of chamber-pots when walking underneath London windows, where 'dropping vaults distil unwholesome dews / Ere the tiles rattle with the smoking shower / And spouts on heedless men their torrents pour'.[11] The effluvia and miasma of the time must have been dreadful. House design reflected this: boot-scrapers were commonly placed at doorways and are often still present at the entrances of Georgian and Victoria-era houses.[12]

Unsurprisingly, diseases were common, and cholera and typhoid struck uncounted numbers over the years. Charles Dickens, who himself undertook long and risky walks at night to ease his chronic insomnia, wrote in *Oliver Twist* that the back alleys of London were 'the filthiest, the strangest, and the most extraordinary of the many localities that are hidden in London'. Dickens also wrote, in *Little Dorrit*, that through 'the heart of the town a deadly sewer ebbed and flowed, in the place of a fine fresh river'. This was the open sewer that was then the vast and great river Thames, devoid of life and horribly polluted. In time, open sewers gave way to closed sewers, and waste handling was prioritised. This was an astonishing engineering achievement of the greatest importance. Engineers designed cities to remove what was an invisible – but real – threat, caused by microscopic pathogens in the atmosphere and in our waste.

Nobody would argue that we should return to the dangerous days of open and filthy sewers in order to save money on construction costs. Many developing countries are now attempting to make this transition to safe waste handling, often with great difficulty; the routine removal of shoes at entrances to private houses in developing countries is one way of handling problems arising from filthy streets. The next frontier for urban design is surely integrating into the design of our villages, towns and cities what we know about mental as well as physical health – and this means making walkability a priority. This is a development in urban design that has been shown to bring a range of major improvements to the life of city dwellers – it's healthier, cleaner, more pleasant and certainly makes our cities more enjoyable and therefore better places to live healthy and happier lives.

Walkability also changes the social character of our cities, of the neighbourhoods within our cities, and of locales within those neighbourhoods. Imagine you have just moved to a new city, one that has extensive urban sprawl and lacks good mass transit. Getting around requires a car. The opportunities for random social interaction are few, because of the design of the transport system. Being in a car militates against easy face-to-face interaction and chanced-upon conversation; every sight of another human being is mediated through glass. By contrast, in a densely packed neighbourhood, where people randomly intersect easily at corners, at cafés, in local shops, people can build a social network quicker and easier.

Besides increasing sociability, making cities walkable has another profound and important effect: it intensifies economic activity. Walkable offices and shops in downtown areas can attract a substantial premium over suburban mall space.[13] There are all sorts of reasons for this, but one straightforward explanation is that when you walk, you have time for experiences in personal consumption that you would otherwise not have if you are in a car. Moreover, the money that you spend stays in the local economy, whereas money spent on fuel for transport, and on the car itself, is funnelled out of the local economy.

In other words, walkable cities have 'aggregation effects'. Walkability makes it easier to have business, social or just accidental meetings and social interactions, because spatial proximity makes all of this possible. It has even been suggested by economists that the more car-bound you are, the less economically productive you are.[14] These benefits – social, economic, health and otherwise – are easily achievable when we are able to do what comes naturally: walk.

In general, it appears that the bigger and richer the city, and, in particular, the higher the rate of economic growth, the faster the inhabitants walk. In 1974 the psychologists Bornstein and Bornstein measured pedestrian walking speeds in fifteen cities and towns in Europe, Asia and North America.[15] They found the pace of life varies with the size of the local population, independent of the particular culture. In general, bigger cities across differing countries and cultures had faster walkers. These experiments have been repeated several times since these original observations, with the attempts to see what it is about living in cities that changes walking speed. Is it that there is a greater reward density in cities (restaurants, seats on trains or buses, or whatever), and the greater competition for those rewards acts to increase the pace of life? The geographers Jim Walmsley and Gareth Lewis suggested in 1989 that 'economizing on time becomes more urgent and life becomes more hurried and harried' because incomes and the cost of living increase, and therefore the value of an urban resident's time in cities is greater.[16] This reflects the idea that increased competition for resources changes our behaviour in many subtle ways. We speed up our walking, unconsciously competing against others seeking the same resources.

Bornstein and Bornstein's paper subsequently became both highly cited and celebrated, but it is unlikely to be the whole story. Presumably, there are city-specific factors resulting in different speeds at which people walk. One could imagine, for example, that in exceptionally densely packed cities like Mumbai, walking speeds might actually slow, simply because of the risk of collision with other pedestrians. One might equally imagine that walking speeds might be quite high in selected places. In extremely hot or cold cities, for instance, people might walk quickly between their car and a building, simply in order to escape the high or low temperatures. Biologists Peter Wirtz and Gregor Ries have argued that Bornstein and Bornstein's results arose because they did not take account of the age or gender composition of the cities and towns that they focused on.[17] In other words, all other things being equal, cities tend to have younger populations, and younger people walk more quickly, on average, than do older people. Similarly, males tend on average to walk somewhat more quickly than females, and any apparent differences between cities and smaller towns might simply reflect the fact that cities have more, younger, faster-walking males than do smaller towns and villages. Wirtz and Ries conducted a series of studies of their own with a much larger sample size and, having taken account of gender and estimated age, concluded that city dwellers in fact did not on average walk faster than town or village dwellers.

But this wasn't the final word – this idea has continued to be tested. In 1999, one of the biggest ever surveys of the pace of life was carried out using data from the largest cities of thirty-one countries.[18] The study examined factors other than population causing differences in the pace of life, and what effect the pace of life has on the well-being of those urban dwellers. There were three different ideas examined that would predict pace. The first was economic vitality: the faster the rate of economic growth, the more vital the economy is, the higher the pace of life might be. The second was that hotter cities on average would tend to have slower walkers; and the third was that countries with relatively individualistic cultures would have a faster pace of life compared to countries with a collectivist culture.

Focusing on cities as diverse as Dublin, Hong Kong and San Salvador, researchers measured walking speed (how fast it took people to walk sixty feet in two differing downtown areas), postal speed (how quickly you could buy a stamp in the major post office), and finally clock accuracy. Other information was gathered from publicly available data sources on climate, economic indicators, measures of individualism, size of population, coronary heart disease, levels of smoking and subjective well-being. These were all combined to create an overall pace-of-life index. Switzerland was measured as having the fastest pace of life, closely followed by Ireland (then in the middle of a major, decade-and-a-half-long economic boom), followed by Germany and Japan. (Italy, England, Sweden, Austria, the Netherlands and Hong Kong completed the top ten, in that order.) Mexico came in last. On the larger global scale, it was shown that Japan and the non-ex-Soviet bloc of Western European countries had the highest pace of life, with people in Ireland clocking the highest individual walking speed. Switzerland lived up to a stereotype, ranking number one for clock accuracy.

One issue that is not entirely resolved in these studies is the relationship between population in terms of numbers, and population density in terms of how it affects speed of walking. Oxford Circus in London at rush hour for example is astonishingly busy, and also consequently difficult to traverse. A couple of streets away, however, people move with relative ease. The new generation of smartphone pedometers and health apps should help resolve this issue – there might an optimal density of people that maximises walking speed, and going much above this number causes walking speeds to slow down.

Let's accept for a moment that the pace of life in cities is faster, and that this has less to do with demographics and more to do with economic vitality and population density. What is it about us as individuals, that when taken from one circumstance – a quiet country town, for example – and placed in another one, a busy city, we increase our walking pace? Researchers have suggested that it could be down to the readily available economic rewards in the city. Suppose we take the (not unreasonable) view that the brain makes calculations of effort and reward. Furthermore, the brain attempts to balance effort and reward, such that effort is minimised and reward is maximised. (Think of this as the brain's evolved response that balances laziness and effort as discussed in Chapter 2.) This raises the question of how rewards are processed by the brain, and in particular the amount of effort that is made in order to attain an outcome. Think of the hip new restaurant down the side street with a no-booking policy. You'll walk more quickly to get there, and perhaps walk more quickly again because you're competing against others who want the same reward – perhaps the best table, or maybe the only remaining table. If you have two choices, you are likely to choose the more rewarding option, or, at least, the one requiring the least effort. We seem to walk faster in the city – presumably because cities have lots of resources and rewards, but we must also compete against others for those rewards. We make differing types of effort in the course of everyday life: whether reaching for something, or walking to something. It seems that across categories of effort, the energy expended to attain a goal rises to some maximum, and then falls away.[19] The neuroscientist, Reza Shadmehr, asks us to imagine you're at an airport. You're standing in the arrivals hall, scanning the faces for a passenger. As you see the successive faces of the arrivals, you spot the person you're waiting for. Now ask yourself, who will you walk faster to in order to greet them: a colleague, or your child? The answer should be obvious. The intrinsic reward resulting from meeting your child would be particularly great. This intrinsic reward, in turn, directly modulates your walking speed, and you make the effort to get to your child faster. We walk faster when there are greater rewards at stake. Effort and the prospect of reward are therefore added together.

Here we have a basis for understanding how our walking speeds vary in differing cities. It is likely that the rich resources in the city mean that people are willing to expend effort to obtain them. At the same time, there is increased competition for those rewards – not only do we have to walk to that nice new restaurant quickly, we must get there more quickly than other people do. In turn, there is a close coupling between the brain systems that manage effort, and the brain systems that estimate likely rewards. The greater the effort, the greater the likely predicted reward stemming from that effort. We will walk more slowly toward things that bring little reward, and vice versa. And this is just what we do in the city: we walk quickly to get to that seat on the train, or that slot in the restaurant, because we are competing with others for the rewards that the city offers.

*

In a busy, bustling city, where we're competing for rewards, a major physical obstacle is of course other people. To avoid collisions, we must be able to quickly and accurately estimate the walking speed of others. When we walk together in pairs or in triples, we fall into step with each other naturally and unconsciously. We moderate our pace so that we can keep to the same speed: we *coordinate* our walking speed with that of others.

A commonly felt aspect of walking in the city can be frustration, or irritation. There are people walking faster than you; people walking slower than you; people walking at you; people walking across you. This irritation has been called 'walking rage' by the psychologist Leon James, who suggests that those 'who impede the flow of progress of others have passive-aggressive sidewalk rage, while those who are impeded by them have active sidewalk rage and walk around with intolerance and disapproval'.[20] A few moments observing pedestrian commuters trying to bypass a drifting crowd of tourists suggests this is a real, if usually self-contained, phenomenon: confrontation is both possible and impossible because of close physical proximity. Social life is only really possible because we continually inhibit our malign impulses. By contrast, travelling in a vehicle allows behind-the-windscreen venting of road rage that can't be safely stated face to face.[21]

Most of the time, we are able to use what sometimes feels like an invisible force field to prevent inadvertently bumping into or making contact with others. We try to avoid them, and they try to avoid us. But how do we manage to negotiate crowded pavements with such ease and grace? What are the mechanics behind it?

To predict where others are likely to move to, we need to rapidly gather information from the movements of others. There are many sources of such information. We might look to the positioning and speed of movement of their feet and legs. Or we might look at their body trunk and the position of their shoulders. Or we might use the overall position and direction of their head, and by looking at the position of their eyes and direction of gaze. When oncoming walkers wear sunglasses, or have their eyes focused downward on a handheld device, your predictions regarding their walking intentions start to break down. More awkward is walking behind someone who drifts laterally. They don't know you are there, of course, and all you have to go on is an inference from the momentary position of their shoulders seen from the rear. And they may in turn be navigating the movement of someone walking towards them, whose eyes in turn you can't see, or whose presence you may be completely unaware of.

To examine the hypothesis that eye gaze is vitally important, you need to test participants with oncoming pedestrians whose eye position varies in systematic and reliable ways. This can be done easily enough now with wearable technology, in particular virtual-reality goggles. Virtual figures approach participants at various speeds as they either look into each other's eyes or avert their gaze (so they can see approximately where the eyes of the person approaching are looking).[22] Moreover, brain-imaging experiments combined with virtual reality show that when the oncomer's eye position gives an unambiguous signal for direction of walking, activity occurs in the brain that readies the body for a course correction. In this little act of navigation there is a rapid and constant flickering between differing mental states. We look to the position of the other person's eyes in order to figure out where they are looking and where it is that they are likely to move to. We then do a rapid course correction in order to minimise contact, and they do the same. If it all works out, then embarrassing shoulder bumps don't occur, and there is no need for a mumbled apology.

These kinds of experiments have revealed which brain networks are likely to be engaged during these forms of social interaction. Two distinct brain areas (the superior temporal sulcus and the fusiform gyrus) are activated strongly, but differentially, during this task. The superior temporal sulcus is a piece of brain tissue about the length of your little finger, located to the side of the head, at about the level of, and forward of, the tip of the ear. The fusiform gyrus is located at about the level of the centre of the ear on the under-surface of the brain, again, on both sides of the head. The superior temporal sulcus is activated strongly in the right-hand hemisphere of the brain, when the avatar walking towards the participant shifts their gaze towards the participant, or averts their gaze. When the avatar and the participant look directly at each other, this causes strong activation in the superior temporal sulcus, whereas an averted gaze causes less activity in the superior temporal sulcus. By contrast, the fusiform gyrus responds approximately equivalently to both mutual and averted gaze. The implication therefore is that the superior temporal sulcus is involved in the processing of social information – the trajectory of walking movements – when that information is conveyed by shifts in the eye position of the person walking toward you. We humans are a social species: we learn from others, including what course corrections to make when we are lost while walking. Having a socially driven input to our cognitive maps allows us, through shared knowledge, to learn from others. Anatomically, the fusiform gyrus provides a major input to the hippocampal formation.[23] Another way of interpreting this is that the input from the fusiform gyrus is the means by which the cognitive map can be modulated by the presence of others.[24]

Our natural patterns of walking are controlled, as we have seen, by mid-level central pattern generators in the spinal cord. These generate your ongoing pattern of walking without much supervision by the brain. However, the central pattern generators are quickly and rapidly interrupted by a command signal from the brain, providing information about the intended movement of the other person, inferred from where they are gazing. There is a rapid top-down intrusion by brain systems concerned with social interaction which interrupts the ongoing work of the central pattern generators. This command signal redirects the activity of the pattern generators: perhaps momentarily causing you to pause your footsteps, rotate your shoulders, pivot, move your bag – all to try and avoid a collision. And the other person is likely – hopefully – doing the same thing: hence, the relative ease with which we can walk in crowds.

Stepping back a little, we see we have two mutually interacting minds in motion: two brains momentarily capturing information about where the eyes of the other brain are gazing, and using that information to make a rapid prediction about where the other person is about to walk to.

Once, late one evening, quite a few years ago, I was walking across the large central concourse of a railway station. Two somewhat raucous and clearly drunk men were walking across the platform as well. I was walking in a diagonal direction, and one deviated away from the other and started to walk towards me. He stared at me in an aggressive fashion, picked up his walking speed, and moved his left shoulder back a little. I could see he was getting ready to hit my left shoulder as hard as he could with his shoulder as we came abreast of each other. As we were about to make contact, I turned my left shoulder away from him and, as I had predicted, he attempted to strike me with his shoulder, except my shoulder was no longer there. He spun in a circle and fell awkwardly to the ground. This was not what he was expecting, because this is not what his brain had predicted. A combination of predicting what he was likely to do, coupled with his staring into my eyes too long and the cocking back of his shoulder, allowed me to avoid what otherwise would have been, perhaps, an unpleasant encounter.

Underlying these capacities is an elaborate neural circuitry which can respond rapidly to the prediction of the behavioural intention of another – and do so in a fraction of a second, while maintaining all of the other vital functions that the brain also has to worry about: breathing, heart rate, digesting, being conscious.

Our brains are deeply social. We can often read what people are about to do just by looking at them. There are lots of possible simple rules and heuristics which we might use to guide our behaviour in crowds. These might be 'Do what the person closest to you is doing' or 'Always move to the centre of the crowd' or 'Run away from the direction that a small group is looking at – but only so long as they are showing fear.' Any or all of these rules might be, and could be, powerful predictors of individual behaviour in crowds, and how the behaviour of individuals coalesces to create the kind of behaviour that emerges when people walk in crowds.

The renowned social psychologist Stanley Milgram devised a simple method to examine behavioural contagion in passers-by.[25] He had people gather and stare upwards at the sky at the corner of a New York City block, and then observed how this simple behaviour affected those around them. He found that, as the group increased in size, the likelihood that other people stopped and stared into the sky alongside them grew much greater. Here, we see again a role for social contagion affecting our individual walking trajectory: we see others doing something indicating something important might be going on. We stop, we gaze, we engage in information pick-up, and we do this rapidly and unconsciously.

Milgram's sky-gazing experiment has now been repeated using modern digital image capture technology, which permits an exactitude of analysis that Milgram could have only dreamed of. Experiments have tracked both the direction of walking and head directions of 3,325 pedestrians, in naturally occurring crowds, either on a crowded street, or on a busy railway-station concourse.[26] The researchers instructed a group of people to stand on one of the busiest streets in Europe

– Oxford Street in London – and stare up at a camera mounted on a building. They then watched the behaviour of 2,822 passers-by, and found, just as Milgram did, that the greater the number of people looking up in the group, the more likely people were to gaze upward too. More than a quarter of passers-by adopted the gaze position and direction of the stimulus group. This is mobile cognition in action in the real world: we see an opportunity for what might be some important information scavenging, and we engage in it quickly, copying others through the action of an elaborate neural machinery of which we are unaware.

They also observed what looked like a cliff-edge effect of some significance: passers-by beyond an approximately two-metre distance did not stop and gaze upwards. Passers-by were most likely to stop and gaze upward if they approached the stimulus group from behind. Upward-gazing, and thereby copying what another person is doing, is not here a simple social conformity effect: the person gazing upward will be completely unaware of someone engaging in upward gazing to their rear. A passer-by's visual attention can be captured, in part at least, by their noticing the position and orientation or direction of the back of the gazer's head.

To a passer-by, looking skyward is relatively benign – you might be looking at a bird, a window display, an ornate piece of architecture. Our cognitive maps must take account of other types of information, however. Not only are we vigilant for sources of reward, we must also rapidly incorporate sources of potential threat and danger into our map of the world. This has been found in separate experiments that involved asking two men to behave in either a 'natural' or a 'suspicious' fashion. In the natural condition, they stood for sixty seconds as if waiting for somebody, while on the concourse of a busy station. In the suspicious condition, they stood, either sketching the layout of the station, or apparently filming the environment with a camera held at waist height. The closer passers-by were, the more likely they were to look at both the suspicious and non-suspicious individuals. In the railway station, people glanced at, and then ignored, the non-suspicious, naturalistic men as they walked by. However, they actively averted their heads while walking past the suspiciously behaving men, perhaps as a form of confrontation avoidance.

People rapidly assimilate potential sources of threat while walking, and quickly correct their walking trajectory to avoid that threat. Our cognitive map of the world is not a static map: it incorporates information offered by others; it logs possible threats and places of refuge, and does so in a dynamic fashion without us being aware of it. In other words, our cognitive maps afford us a flexible way of interacting with the world while we are moving through it.

*

What gives our cities their vitality, attraction, upsides and downsides? How do walkable cities acquire their sociable character? What is it about cities that attracts people to live in them, despite their downsides? There is a density of social interaction afforded by crowding people into compact urban spaces. Cities do change you – and you won't even know it. The life you have, the way you think about the future and the past will change if you are living in the city. Greater competition for resources subtly alters our behaviour in ways that we are hardly aware of, even the pace at which we walk. People leave the city to enjoy a slower pace of life, and go to the city in order to be stimulated by it.

I have often walked the towns and cities of Italy and have always been struck by that wonderful tradition of the *passeggiata* – the sociable, evening stroll along the streets, greeting and chatting with neighbours and friends. This habit of walking the city sums up everything good about urban walking and shows how entirely achievable this is as a daily activity, if only our cities were built for it. I offer the acronym EASE to assist our city designers. To enable the *passeggiata*, our cities should be *easy* (to walk); *accessible* (to all); *safe* (for everyone); and *enjoyable* (for all). The steps to make and remake our cities walkable are straightforward. We just need to take them.

# 6.
# A BALM FOR BODY AND BRAIN

Moving about the world is an essential part of the experience of being human. Movement, and, most especially lots of regular walking, is good – indeed, great – for body and brain. In this chapter though I want to go far beyond these simple and relatively uncontroversial contentions and investigate the potential wider benefits of walking – to ask how walking affects mood, mental health and brain function? Regular walkers (myself included) claim that, deprived of the opportunity to walk for even a few days, we feel sluggish and tired, and often a little bit down, and that the self-administered cure is simple – to go out for a good walk. Thrillingly, there is now an emerging body of science that supports this anecdotal feeling, and which indicates that walking, especially in regular doses, often in nature, does actually improve how we feel. Think of all those blustery, rainy, long walks that at the time might have felt arduous, but at the end left you feeling elated. A good walk boosts how you feel, and much more besides.

Hippocrates famously claimed that 'walking is the best medicine'. Yet in our modern world, most of us spend all day indoors sitting down, which can have terrible consequences for our health and well-being. We spend less time outdoors than ever before. One major study in the USA showed that people spent 87% of their time in the artificial environment of offices, houses, shops and other buildings.[1] Some have even claimed (only somewhat exaggeratedly, in my view), that 'sitting is the new smoking'. The sentiment behind this statement is straightforward: our bodies are built for regular movement, and profit from it. Sedentary life is fundamentally unhealthy, leading to decline in muscle volume and strength.[2] Moreover, long periods of inactivity produce not dissimilar changes in the brain.

One interesting study has recently found that lack of activity is even associated with personality change, and by this I mean change for the worse.[3] Overall, lower levels of physical activity were associated with changes in three of the 'Big Five' factors of personality (these are *openness, conscientiousness, extraversion, agreeableness* and *neuroticism*, easy to remember as OCEAN).[4] Lower levels of physical activity were associated with declines in openness, extraversion and agreeableness, suggesting a 'detrimental' pattern of long-term personality change. Even minimal levels of activity were found to have a moderating effect on personality change. Those individuals who were the most inactive were the ones most likely to show these negative personality changes.

The pathway channelling these negative changes is unclear, but is likely to involve the usual increases in illness and lack of well-being associated with prolonged inactivity; the limitations to activities of daily living associated with inactivity; changes in general cognitive function; and perhaps even changes in mood. Given what we know, it's highly likely that a simple behavioural change – lots of walking – would be a viable way of reversing negative changes in personality resulting from a sessile life.

Standing leads to immediate changes in blood pressure, blood flow around the body, and the rate at which we consume energy and generate heat (our 'metabolic rate'). Walking entrains changes across widespread brain and body systems, from the production of new molecules all the way to behaviour. Regular, up-tempo, walking is a simple and straightforward way of exercising the heart, and this in turn provides great benefits for the head–heart axis, because about 20% of the output of the heart is directed toward the oxygen-hungry and energy-hungry brain. Similar effects occur in the gut, which is also oxygen-hungry and energy-hungry. The cure is right in front of us: to get up and walk.

*

Walking is one thing: where we walk is quite another.

As we explored in the last chapter, as more and more of us live in towns and cities, green spaces will only become more essential for our well-being. Building design, especially in northerly and more inclement regions, has, in some respects, historically taken account of this fact. Cloisters in university buildings, monasteries and other locations, allow people to walk outdoors while protected from the elements. Cloisters are sometimes referred to by their ritual and processional purpose – *deambulatorium, obambulatorium, ambitus* – the solemn, Latin descriptors of the architectural elements of a monastery, all derived from the root '*ambio*' – 'I walk in a circle'.[5] They are also often called, appropriately enough, ambulatories. And, of course, cloisters are usually constructed around a garden – ensuring a tamed element of nature is at the centre of the walk.

Walled gardens, dating from early times, are another way of bringing tamed nature within a building perimeter, yet allowing safe walking outdoors. In *The Decameron*, Giovanni Boccaccio writes of one such garden that 'its outer edges and through the centre ran wide walks as straight as arrows, covered with pergolas of vines which gave every sign of bearing plenty of grapes that year ... The sides of these walks were almost closed in with jasmine and red and white roses, so that it was possible to walk in the garden in a perfumed and delicious shade, untouched by the sun, not only in the early morning, but when the sun was high in the sky.'[6] Modern building design incorporating cloisters, awnings, courtyards and other features could make outdoor walking and exposure to nature easily achievable. Similarly, indoor walks around nature-bearing and displaying atriums could provide people with this feeling of connectedness to the natural world. Views from windows which provide glimpses of the sky and of trees could also enhance well-being markedly.

Yet this need in our lives for time outdoors and connection with nature is something we consistently seem to underestimate. This has been shown clearly by a study conducted in Ottawa in Canada. Ottawa is subject to weather extremes, with summer temperatures exceeding 30°C, and winter temperatures below −20°C. A substantial fraction of the large campus of Carleton University in Ottawa is connected via a system of extended underground tunnels to allow walking during weather extremes. Experimental psychologists examined the effect of people undertaking walks where they are exposed to nature versus walking in an enclosed environment by using this network of tunnels.[7] They asked 150 participants to walk the same distance between two locations on the campus: either through an underground tunnel or outside along a riverbank in a 'naturised' urban space, with plentiful trees, plants and other features of the natural environment.

Prior to starting, participants were asked to state how they were currently feeling, and then to estimate how they would feel after the seventeen-minute walk outdoors compared to the same walk indoors in the tunnels (using a rating scale). The results were clear: all participants substantially underestimated how the walk outdoors would make them feel relative to the walk indoors. The effect on mood of the walk through the naturised urban setting was compelling. There was an improvement in the individuals' self-rated mood scores of about one-third on average, relative to individuals who undertook a walk indoors. (This study also demonstrates a persistent problem with how we humans understand what affects our feelings: we are bad at forecasting how any activity is likely to make us feel – known as 'affective forecasting'.[8])

But why does walkable green space matter so much for our well-being? What is it about nature that makes us feel better? Walking in the woods is something we humans have done since time immemorial. Some cultures venerate this experience: the Japanese, for example, have the glorious tradition of 'forest bathing' (*shinrin-yoku*): the practice of absorptive, enveloping walking in deep forests for the soothing properties of being connected to, and fully immersed in, the sights, sounds and feel of nature.[9] Forest bathing is an important manifestation of something that appears to be a universal in human experience – a veneration of nature as foundational to our lives, from early pantheistic theories which imagine that spirits inhabit trees, woodland brooks, stones, and the like, through religions that worship the Earth Mother or deities (like the Inca goddess Pachamama), to the present-day idea of 'Gaia' – scientist James Lovelock's contention that we should regard the planet and all life on earth as a single, self-regulating ecosystem.[10]

Certainly, many take the view that we need to care for nature and that nature as a source of well-being is central in our lives. There is also the great concern that human activity is having perhaps irreversible, and certainly malign, effects – from species hunted to extinction to contamination of water courses and seas with plastics, effluvia and toxic materials, to human effects on the climate of the planet itself.

Scientific evidence also backs up our intuitive feeling that regular exposure to nature and the natural world has effects on human health and welfare which are positive, measurable and enduring, and should be thought of as being akin to the provision of clean water, reliable electricity, public vaccination or public hospitals. The evidence to support this can be found by measuring people's stress levels before, during and after their interactions with nature. The stress hormone cortisol is at the core of our 'fight, flight or freeze' response. Cortisol is released in response to the presence of stressors, with potentially positive and negative effects. In the short term, it is adaptive, mobilising resources to help overcome stress. However, the chronic and sustained release of cortisol leads to a whole variety of problems, from stiffening of our arteries and veins, to malign effects on our mood and memory. One study in a very deprived area of Dundee looked at how the amount of green space in a neighbourhood might affect the levels of stress in residents of that neighbourhood.[11] This was measured both by perceived levels of stress (in other words by self-report, how residents thought they were feeling) and by measuring levels of cortisol, which can be readily measured in both saliva and in the blood. The concentration of cortisol in our saliva varies across the course of the day, peaking in the early morning, and decreasing towards the end of the day. People experiencing high levels of stress do not show this downward shift as night-time approaches. In the Dundee study, researchers found that this diurnal decrease is absent, or at least relatively absent, in a deprived population who do not have regular access to and use of green spaces in their urban environment.[12] Finding a correlation of this type is suggestive, and matches to a similar body of evidence which suggests that exposure to nature may have important effects on human health and psychological well-being.

However, we should consider *how* people use the available green space. Do they visit it regularly? Do they use it for social walking, to walk the dog, to supervise children playing? This is where larger-scale studies are required, and preferably studies that attempt to randomise treatment conditions, so that some degree of causality can be figured out. Are your stress levels lower because you are exposed to nature, or is there some other factor? It may turn out, for example, that extended experience of wild nature, involving long periods trekking or walking, might be a viable treatment for depression (at least in its milder forms) and, perhaps, even other stress- and anxiety-related conditions.[13] Large–scale trials testing this idea have not been conducted however.

Getting at whether or not exposure to nature has a causal effect in making you feel better – that exposure to nature generates positive mood – requires studies that vary the dose of nature you are exposed to: does it take a little, or a lot, and how often? The effects may be strong, weak, subtle, or indeed non-existent; the risk of fooling yourself into thinking there is an effect when, in fact, there is none, is high.

'Attention restoration theory' is the idea that the natural environment has profound restorative effects on our well-being, and that the human experience of the natural world markedly assists in maintaining and fostering a strong sense of subjective well-being. According to psychologists, a natural environment should have three critical elements to be fully restorative: it should give you the sense of being removed from your normal life and surroundings; it should contain visual elements and sensory elements that are fascinating in some way; and it should be expansive – it should have some degree of extension.[14] The increasing pressures of modern life tend to increase mental fatigue, but restorative experiences in nature might decrease it. This restorative effect is best mediated through a connection to natural environments because they play an essential role in normal human functioning.

In a study that involved 4,255 participants in the UK researchers investigated this phenomenon of 'restoration', defined as feelings of calm, relaxation, revitalisation and refreshment as the result of visiting a natural environment in the previous week.[15] The recalled restoration from a visit to nature was very high, with an average score of four on a scale of one to five. There was a hierarchy of locations, with coastal environments providing the greatest feeling of restoration, followed by the rural countryside, with urban green spaces coming in third. This hierarchy should perhaps be treated with some degree of caution, though – it is derived from an overall average, and many town parks were just as restorative as the open countryside. A majority of the highest socio-economic group (53%) visited nature in the previous week, whereas only a minority (31%) of the lowest socio-economic status group did so. The higher socio-economic group, of course, will have, on average, better education, health status, access to nutrition and the like.

What's clear is that park design is a vital factor: the extent to which a park is usable, accessible, and facilitates different meaningful activities is the driver of park usage. The differences in feelings of restoration found between time spent in these various environments – the coast, the rural environment, urban parks – were not especially vast, and the study did not control for the activity that you could undertake in the differing areas. Urban green spaces can be used for tending to vegetables, as in an urban allotment; walking the dog, as in an urban park; or playing sports in urban sports fields. Easy access to nature is very important to individuals, to families, to social groups and to society at large, and well-designed urban green spaces can substitute for, or mimic in important ways, the effects of being in the countryside. Parks, for example, might allow wilderness areas supporting urban wildlife, insects and birds, as opposed to carefully mown and tended grasses. Equally, the trails inscribed in these parks should, to the greatest extent possible, follow the undulations of the environment and of people's 'desire paths'.

It's also been shown that the positive effect on mood after spending time in nature applies to a range of people of different ages, both male and female, across the globe. Perhaps more importantly, the impact of exposure to nature is comparable to other factors affecting individual happiness, including personal income levels, level of education, degree of religiosity, marital status, volunteering and physical attractiveness.

It may not be possible to do much about one's personal income, or indeed one's perceived physical attractiveness, but getting out and going for a walk is something that we can all easily do. Because the evidence suggests that activity in nature has a long-lasting impact on our happiness and well-being, we should be encouraging our populations to regularly, habitually, walk in nature, even if they only have access to city parks.

*

Does the fact that walking can demonstrably lift your mood mean that it could assist, in some way, with the experience of depression, perhaps as a protective factor, a sort of 'behavioural inoculation'?

There's an important difference between a mild and transient down-in-the-dumps feeling, to which we are all subject, and major depressive disorder (MDD). MDD has been defined as 'a depressed mood or a loss of interest or pleasure in daily activities consistently for at least a two-week period. This mood must represent a change from the person's normal mood; social, occupational, educational or other important functioning must also be negatively impaired by the change in mood.'[16] The World Health Organisation regards MDD as one of the greatest hazards to health and well-being over the coming decades, with the lifetime risk of suffering an episode of MDD at up to 15% in Western populations.[17] This form of depression usually requires psychiatric care, cognitive behavioural therapy and/or drug treatment.

There has been widespread interest in physical activity or exercise as an intervention for this form of severe and disabling psychiatric disorder. Some large-scale trials have tested if physical activity (walking being a particularly common intervention) relieves MDD. As we've seen, walking and other forms of physical activity do have cumulative, consistent and positive effects on us all, and the same is true for sufferers of MDD, and at a level that can be comparable to drug and cognitive-behavioural intervention treatments.[18]

Studies into whether we could see walking almost as a kind of immunisation against depression are, however, problematic. A randomised control trial methodology is more or less impossible to conduct *ante hoc*. By this, I mean one cannot select a large sample of the population, then subject a fraction of them to life stressors designed to induce depression, and then look at rates of depression depending on the dose of walking that has been administered prior to inducing depression. Rather, answering a question like this involves recruiting very large samples ('cohorts') and tracking their levels of physical activity. Then you must estimate the rates of occurrence of depression in that population, and show that the likelihood of major depressive disorder is lower the more active a person is.

These latter types of studies are in principle no more difficult than something like large-scale drug trials. One major study led by the psychiatrist and epidemiologist Samuel Harvey of the University of New South Wales focused on a cohort of 33,908 adults, selected for the absence of any common mental disorders or illnesses, and for the absence of any physical diseases which might limit movement. These 'healthy' participants were tracked for eleven years.[19] The study attempted to show whether or not exercise provides any protection against the onset of depression in participants who have not been previously depressed; it also attempted to show what level or dose of exercise is required, and the extent to which any particular mechanism underlying any protective effect could be elucidated.

Self-reports of exercise, and independent validation of that self-report by measurement of oxygen uptake were undertaken. Other measures included demography, socio-economic factors, smoking and alcohol, body mass index, social support and the like. Overall, the study found that engaging in any physical activity at all was positively associated with a reduction in the likelihood of depression (but, interestingly, not anxiety). The protective effect for exercise against depression occurred even for very low levels of exercise: infrequent walking, approximately once a week or so. The overall conclusion suggests that, assuming that activity underlies the observed reduction in cases of depression, approximately 12% of future cases of depression could be prevented if 'all participants had engaged in at least one hour of physical activity each week'. These are relatively modest changes to activity levels, but the positive effect they have is far greater.

The study conducted by Harvey and colleagues is to date the largest on record, with walking and swimming the form of exercise that most of their participants undertook. They conclude that as the intensity of exercise does not appear to be important, 'It may be that the most effective public health measures are those that encourage and facilitate increased levels of everyday activities, such as walking or cycling.' In other words, designing our environments to nudge us to take more activity might be the solution. This study does not argue in any way against the importance of exercise for cardiovascular function. It does argue, however, that increasing levels of activity over the baseline of no activity at all have an important effect on the emergence of subsequent episodes of depression.

The public health physician Gregory Simon, on reviewing these findings, concluded that 'exercise is a safe and moderately effective broad spectrum antidepressant prescription ... for prevention and treatment across the spectrum of depression severity' – an important policy recommendation for the years to come, made in a leading journal of psychiatry.[20]

The question of the relationship between walking and mood is a tricky one. It is genuinely difficult to get clean answers from any experiment. Are people who walk more generally a little happier or somewhat less likely to succumb to depression because walking elevates mood? Or because you are happier, are you as a consequence more active, and do you therefore walk more? Which is cause, which is effect? Or is there a a positive feedback loop here? Once the initial inertia to start walking is overcome, walking in turn may increase feelings of well-being and happiness, which in turn increases the willingness to engage in walking.

The studies we have, in humans at least, are largely correlative and indicative. While they point in the direction of a positive relationship between walking and mood, what is less clear is whether or not an active programme of walking enhances mood more generally, either while walking, or in the periods of time after walking. Furthermore, the issue of the relationship between the dose of walking, and mood, is also problematic. One important source of evidence is to look at the relationship between mood and walking either through self-reports (which we know can be very unreliable), or through automated capture of the number of walking steps combined with moment-to-moment, self-assessed, current mood.

Another alternative is to use exercise or walking as an intervention in individuals who are assessed by a clinical psychologist or psychiatrist as suffering from clinical levels of depression. To be meaningful, this requires conducting appropriately designed and controlled clinical trials – not an easy task. It might be that if those suffering from MDD substantially increased the number of steps they undertake every day this might (and I emphasise might) act as a kind of behavioural antidepressant. It may also be the case that walking serves to lift what is referred to as transient dysphoria – that feeling of being down in the dumps or under stress that we all suffer from time to time. In this context, the phrase 'walk it off' may mean something. In the case of major depression, though, it may not.[21]

An unappreciated way walking boosts mood is through the pleasure derived from resting after extended physical exertion – in a warm bath, or simply sitting in a comfortable chair.[22] The great British philosopher Bertrand Russell remarked that 'I used, when I was younger, to take my holidays walking. I would cover twenty-five miles a day, and when the evening came I had no need of anything to keep me from boredom, since the delight of sitting amply sufficed.'[23] While the relationships between mood and walking are not straightforward, the emerging findings do suggest that regular walking has acute and chronic effects on mood, boosting feelings of well-being both in the moment and long-term.

*

Walking and other forms of aerobic exercise can also have a profound effect on learning and memory. A vast literature of thousands of research papers extends from describing underlying molecular changes in the brain resulting from learning, to the effect of exercise on memory and cognition in the elderly.[24]

In 1949 the great Canadian psychologist Donald Hebb suggested that the key to how memories become inscribed in the brain is the zone of contact between one brain cell and another (the 'synapse').[25] Hebb predicted that synapses had to be plastic, meaning they could be remodelled as a result of experience: the more frequently one cell or neuron participated in the firing of another cell, the more likely that the first cell would more easily fire the second cell, compared to another cell attempting to fire the first cell. In this way, there would be a pattern of 'co-activity' present between cells. The plasticity of synapses could form the basis of how memories might be written into the fabric of the brain.

This far-reaching idea gave rise to the slogan 'cells that fire together, wire together'. For 'wiring together' to occur implies there must be structural changes or remodelling happening at the synapse. This means there must be molecules present, or arising as a result of activity, supporting or facilitating these changes. One of the key molecules is brain-derived neurotrophic factor, or BDNF, which might be thought of as a kind of a molecular fertiliser produced within the brain, because it supports structural remodelling and growth of synapses after learning.[26] What we have learned recently is that a simple and straightforward way to raise BDNF levels in the brain, and particularly in the hippocampal formation, is aerobic exercise.

Many experiments have demonstrated reliably that giving rats and mice the opportunity to exercise in running wheels increases significantly and reliably the level of BDNF in their brains.[27] These animals are subsequently able to learn maze and other tasks more easily than their littermate controls who do not have access to exercise wheels. Moreover, if the increase in BDNF is prevented by specialised drugs that block this increase, the exercise-induced advantage in memory disappears.

Many labs have now shown that aerobic exercise also increases the creation of new brain cells in the mature adult brain. It used to be thought that after birth no more brain cells were created. The scattered findings in the scientific literature in the decades preceding the 1990s, demonstrating that new brain cells can arise in the adult brain, were largely overlooked. Eventually, however, the weight of data became so great that the previous orthodoxy was overthrown. New brain cells are, however, only made in a few locations in the brain (a process known as 'neurogenesis'). One especially important location is in a part of the hippocampal formation known as the dentate gyrus.

Blocking the creation of new brain cells can cause disturbances in learning and memory. Moreover, animals in whom neurogenesis is blocked – for example, as a result of behavioural stress – show a depressive-like symptomatology: reduced levels of grooming and self-care, and reduced attempts to escape stressful situations (such as being placed in a small bath of water); they also show deficits in maze and reward learning and memory. Other work demonstrates that activity in the hippocampal formation arising from effortful learning or aerobic exercise is required for the incorporation of these new brain cells into the hippocampus itself. BDNF provides a core, supporting molecular mechanism for learning and memory in the brain; BDNF is produced because of behaviour, including aerobic exercise, and in turn, the production of BDNF boosts learning and memory very effectively. BDNF also provides the brain with a remarkable degree of resilience, including resistance to ageing, and damage arising from trauma or infection.

Demonstrating these changes in animals is relatively straightforward – in a post-mortem examination we are able to measure changes in an animal's brain after an experiment. For obvious reasons this is not the case in experiments with humans, although both direct and proxy measurements can still be taken from saliva and blood, as well as responses to a variety of cognitive and other tasks (using fMRI we can measure many changes in the brain's structure and function too).

Our current consensus model of learning and memory suggests that a variety of brain regions must act in concert to support normal learning and memory: the instruments must play in the correct sequence in order for a memory to be laid down in the brain. As we have seen earlier, the hippocampal formation acts as a critical hub in a distributed brain network, involving a wide variety of cortical regions (remember our simple hand model of the brain in Chapter 3) as well as certain subcortical regions, especially the rostral and anterior thalamus.[28] We now take a 'network-centric' view of how the brain supports particular functions. No longer do we talk of the area for x – for language, seeing, feeling, movement. Instead, we have a view of the brain that suggests that the patterns of interactions between differing brain regions are critical for supporting functions like learning and memory, as well as language or vision or hearing.

Taking regular aerobic exercise enhances blood flow through the brain, as well as making a marked difference to the structure and function of the brain. Aerobic exercise facilitates the brain in populating one particular region, critical to learning and memory, with new brain cells. Moreover, aerobic exercise supports the widespread production of key molecules that act to keep the brain in good working condition. Running as an aerobic exercise is a powerful method of inducing these changes, but running comes with disadvantages (special shoes, preparation, changing clothing, showering). More than these small inconveniences, the risk of injury rises with the distance run, whereas injury risk stays approximately the same with walking, irrespective of distance walked.[29] One small-scale study concludes there is lowered injury risk for walkers compared with joggers.[30] To get the maximum health benefits from walking, speed should be consistently high over a reasonable distance – say consistently over 5 or 5.5. kph, sustained for at least thirty minutes, at least four or five times per week.[31] Walking offers the supreme advantage of needing nothing more than a decent pair of shoes, perhaps a raincoat, and little else. If undertaken in regular doses during the day, it provides the small, cumulative and significant positive changes for lung, heart and especially brain health. Brain health here is to be understood in its most general sense. Just as the purpose of the heart is to pump blood, and the purpose of the lungs is to allow respiration, the purpose of the brain is to support the full range of things we do during the course of our lives, from thinking, to remembering, to problem-solving, to planning, to the regulation of our moods, and a myriad of other things. Regular walking promises, when conducted at a reasonable tempo, to be the shortcut that boosts brain function across the board.

Getting up on your feet and moving involves your whole body: your hands and arms to pull you out of your seat and balance your moving torso; hips to control gait; heart to supply blood and oxygen; a whole gamut of activity. To what extent are changes in the brain driven by changes in the body? BDNF supports a wide variety of important changes in the structure, function and connectivity of brain cells. Brain cells need a constant supply of nutrients and oxygen: they grow and develop within a brain that needs the blood provided by the pumping heart. Blood vessels therefore must evolve in response to the demands for oxygen and nutrients made by plastic, living brain cells.

Another molecule which is stimulated by exercise plays a pivotal role in brain plasticity. Vascular endothelial growth factor (VEGF) encourages the growth of the vasculature, the network of tiny blood vessels that carry oxygen and nutrients to individual brain cells. Strengthening the connections between brain cells, the building and maintenance of the networks that sustain thought and behaviour, is not possible without constant traffic from outside the brain into the brain itself. A brain, like a city, needs a transport infrastructure to ensure a constant supply of nutrition and raw materials.

What might this traffic consist of for our brains? Blood, oxygen, nutrients and other molecules made within the body are carried in; detritus, waste products and molecules made within the brain are carried out. The traffic is two-way, implying that activity in the brain is affected, perhaps even determined, by molecules made in the body.

How do these molecules produced by the body in response to movement act within the brain? We now have the rudiments of an answer to this question. Researchers at the University of California, San Diego, investigated how neurogenesis is triggered in the hippocampal formation.[32] Using genetically modified mice, researchers showed that a molecule made by actively working muscle (with the unlovely name of 'skeletal myofiber vascular endothelial growth factor', or smVEGF) is required for exercise-induced neurogenesis in the hippocampal formation. Walking works muscles in your arms, diaphragm, abdomen and neck, as well as your legs – so there are lots of places where this molecule is produced, and can diffuse through the blood supply into your brain.

Think about the implications of this for a moment: regular movement of the body causes the release of smVEGF, which then circulates in the bloodstream, and is thereby carried into the brain. Once there, it stimulates the growth of new blood vessels, and in turn, supports the production of new brain cells. That the brain is permeable to outside influences from activity in the body is a central theme in this book, and an idea that is supported by findings that there is positive feedback and positive crosstalk between the activity of skeletal muscle and the brain. The necessary conclusion is that walking, or locomotion, when conducted regularly and reliably, generates molecules (in this case, smVEGF) in the musculature, which in turn stimulate the positive changes found in the brain as a result of regular exercise.

'Use it or lose it' is a primary rule that muscle cells obey, and the same is true of brain cells. The body cannot afford to waste energy building, supporting and managing muscle cells or other cells that do not contribute to the overall life of the body. The best signal that the body gets to tell it that muscle is required is when the muscle is placed under regular (though moderate) stress and strain. If an organ is working, it is being used, and if it is being used, it needs to be maintained. The muscles of people who are relatively sedentary for long periods of time start to change, with fatty deposits being laid down in their inactive muscle. These deposits have been found in the muscles of astronauts, the obese and the elderly, who do not get enough of the type of exercise that places positive strain on the muscles of the body.

We now have a precise picture of how rapidly inactivity causes these changes in muscle. By using a technique known as 'dry immersion', which immobilises the muscles and reduces the load they usually bear, we are able to see these changes as they evolve in real time.[33] Volunteers in this study lay on a waterbed designed to support their bodies completely, front and back. Imagine a waterbed shaped something like an envelope into which you insert yourself and which supports and immobilises you. Their body temperatures were kept at a controlled and consistent level through the circulating water, in order that they did not overheat. Participants lay in this microgravity position for three days, except for short toilet breaks, during which many of the markers associated with changes in muscle arising from inactivity were measured.

The results of even this short period of inactivity were startling. The 'viscoelasticity' of thigh muscle decreased by 9% on average (viscoelasticity is a little like the combined properties of rubber and water – it can stretch and flow), indicating that it was less functional because of disuse. Moreover, experiments measuring the ease with which participants contracted and relaxed their thigh muscles showed a uniform decrease in performance. Muscle biopsies (these were brave volunteers, allowing muscle to be extracted and measured via hypodermic syringe) showed significant atrophy (or shrinkage) as a result of just three days of immobility. These changes were also visible on MRI scans of the thigh muscle.

Being sedentary is bad for you, even if you are young and fit: your muscles will decrease in volume, quickly and easily, if they are unused. Moreover, loss of muscle mass is also associated with a loss of the production of molecules important for supporting new brain cells in the few regions of the brain that continue to produce new brain cells through life. As your muscles deteriorate, your brain is also deteriorating. Other malign changes occur too – in personality, in mood, in the very structure of the brain. And yet, we have this wonderful, in-built correction mechanism, a form of self-administered medicine, one without adverse events: movement.

# 7.
# CREATIVE WALKING

I love walking for all sorts of reasons – but near the top of the list is that I find it the best way to clear the clamour of the day from my head. Walking gives me the freedom to think things through; to have a quiet dialogue with myself about how to solve a problem. The problems may be mundane, but are nonetheless important to me. And I'm not alone. Since antiquity it has been recognised that a good walk is an excellent way to think problems through. The school of peripatetic philosophy in ancient Greece was famous for conducting its teaching largely on foot – indeed the root of its name means 'walking up and down'. The philosopher Friedrich Nietzsche went so far as to say that 'Only thoughts reached by walking have value.'[1] In a similar spirit, the writer Henry David Thoreau observed that

> the moment my legs begin to move, my thoughts begin to flow, as if I had given vent to the stream at the lower end and consequently new fountains flowed into it at

> the upper. A thousand rills which have their rise in the sources of thought burst forth and fertilize my brain ... Only while we are in action is the circulation perfect. The writing which consists with habitual sitting is mechanical, wooden, dull to read.[2]

These are important insights, suggestive of a vital relationship between movement of the body and the flow of thinking. Moreover, they emphasise movement's relationship to creative work. Many have written eloquently on how walking brings a clarity of thinking, of creativity, of mood. William Wordsworth composed his poem 'Tintern Abbey' while rambling, saying 'I began it upon leaving Tintern ... and concluded ... after a ramble of four or five days ... Not a line of it was altered, and not any part of it written down till I reached Bristol.'[3] About a century and a half ago, the Danish philosopher Søren Kierkegaard wrote that 'Every day I walk myself into a state of well-being and walk away from every illness. I have walked myself into my best thoughts, and I know of no thought so burdensome that one cannot walk away from it.' The writings and routines of artists and philosophers through the ages are peppered with remarks like this.[4] So, it is not a twenty-first-century revelation that walking, one of the most commonplace of wonders, that affects so much of what we do, directly and indirectly, can free our minds to reach their most creative states. But how and why that is so is the area in which some of today's most exciting cutting-edge research is being conducted.

*

In my case, walking is a useful aid and spur to writing. I have a book to write. I plan things out carefully: chapters, headings, themes. I read, make notes and bullet points, get my Dictaphone, and walk and dictate. A half-hour of planned walking and talking turns into an hour or more. I talk far beyond my notes. I do this repeatedly. Over the weeks and months, 50,000–60,000 words might come together – the basis of a book, something to be revised, modified, edited. I need to speak in shorter sentences, and use fewer subclauses. But whatever the imperfections, walking can provide a certain fluency for writing and dictating, and so long as you don't mind curious looks, it works. Moreover, I find going for a walk before writing also helps the subsequent work, organising my thoughts into some form of order. Bertrand Russell was an astonishingly active walker, and his walking life is dotted through his autobiography.[5] His friend, the actor and dramatist Miles Malleson, wrote that 'Every morning Bertie would go for an hour's walk by himself, composing and thinking out his work for that day. He would then come back and write for the rest of the morning, smoothly, easily and without a single correction.'[6] Walking with some cognitive focus readies you for writing.

I especially like to walk the pathways around Victoria Hill, part of Killiney Hill Park in south County Dublin, close to where I live. The views from the obelisk at the top of Killiney Hill are spectacular and reminiscent in part of the Bay of Naples, hence the Italian names of roads in the area – Nerano, Sorrento, Vico, Marino. The sea winds and gusts can be a problem, though – hold your notes securely. And the reason walking works for me as a spur to thinking, and thence to writing, is that it fosters a simple, quick and efficient flicking between different mental states. Walking allows me to zoom in on a thought, and then zoom out again, placing it in the context of other things.

Here's a simple task: count the number of occurrences of the letter 'e' in the previous paragraph. Finished? Good. Now spend a few seconds thinking about some of the good moments of your life. What you'll find is that you can switch relatively easily between a focus on details (counting the 'e's) and the big-picture events (Where was I? Who was I with? Was I happy?). This little exercise demonstrates how the brain has two essential work modes: an active, executive mode, and a default mode.[7] The active mode involves focused attention and processing details. The default mode involves mind-wandering, the repeated interrogation of autobiographical memory, and a focus of attention away from the immediate environment.

We spend huge amounts of time mind-wandering. Experience-sampling studies – which use mobile phones to ping individuals during the course of the day to ask them what they are doing – show that humans spend substantial amounts of time in this default state, with some estimates suggesting as long as ten to fifteen minutes every hour as we drift in and out of engagement with our environment.[8] But mind-wandering is not mere idleness or time-wasting, at least by the common understanding of the term: rather, it is a necessary part of mental housekeeping, allowing us to integrate our past, present and future, interrogate our social lives, and create a large-scale personal narrative. If mind-wandering is idleness, it is a peculiar and active form of idleness – we are behaviourally quiescent, but mentally vigorous. Children are often told by teachers not to stare out of windows and to pay attention. But it seems that generations of teachers might have it wrong: task-focused attention and mind-wandering are two sides of the same mental coin. We focus on tasks in order to perform them, and we focus away from those tasks to gather the resources to allow us to solve them when they are difficult, or simply to integrate the information that we have learned during the focus on the task. This flickering between states is what allows us to live a productive and creative life; a variety of experiments show that active mind-wandering facilitates subsequent creative problem-solving, with participants who have mind-wandered providing both more and more creative solutions compared with those who have not.[9]

Walking is, paradoxically, a form of active idleness, and it facilitates engaged mind-wandering. It is directed action and focused, but walking allows the mind to wander easily, with thoughts ranging over the day to come, the day passed, the year ahead, the decade passed, opportunities taken or missed. In his great novel *Ulysses*, James Joyce caught this idea well: 'We walk through ourselves, meeting robbers, ghosts, giants, old men, young men, wives, widows, brothers-in-love, but always meeting ourselves.' While walking, you can talk quietly with yourself and aloud with others; or simply listen to music, audiobooks or podcasts. The advantage of walking in company is that it facilitates information exchange, and the integration of that information with your own memories, thoughts and feelings.

At its most simple level, the reason why walking especially enables this flickering between mental states may be structural, founded upon certain key brain regions. The brain system that is active when we access memory revolves around the hippocampal formation, and the many structures connected with it – the extended hippocampal formation. Importantly, the extended hippocampal formation is also active when you are walking, running, navigating an environment. This one brain network supports (at least) two interrelated functions: episodic memory on the one hand, and spatial navigation on the other. One possibility as to why the flickering between these two modes of thinking lies at the core of creativity arises from the notion that in order to create something new, you must combine ideas in some form of novel association. Mind-wandering allows the collision of ideas, whilst mind-focusing allows you to test whether it is nonsensical or interesting and new. The more we look, the more we find that the hippocampus plays a central role in both these activities, facilitating richly productive cross-pollination in the brain. We can even think of mind-wandering as a form of 'divergent thinking', where you muse far beyond the normal constraints of a problem and its possible solutions.

While in the default-mode state we are generally engaged in big-picture, autobiographical thinking about our pasts and possible futures. We spend time thinking about others too – we engage in *social cognition* in the absence of others. We may also even engage in fantasy. Perhaps another way of thinking about the default mode network is that it is involved in the construction of stories and narratives about ourselves, and the wider world in which we live. As a result, some theorists have gone so far as to suggest that the activity of the default-mode network constructs the core of selfhood, because it is so active and engaged during acts of autobiographical recall.[10]

Experiments have shown that when we watch films, read stories, tell stories or have stories read to us, it is the default mode network that is active. As we've seen, the theory that creativity can emerge from flickering between default mode activity and task-positive activity is attractive. But an even more attractive idea, which is what one important recent study concludes, is that creativity emerges when both the task-positive network and the default network are engaged *simultaneously*. Think of this as trying to see the forest and the trees simultaneously: focusing on the detail, while focusing on the whole.[11]

Psychology and neuroscience have been slow to recognise the benefits of walking as a spur to creative thinking. The Nobel Prize-winning psychologist and economist Daniel Kahneman, however, is one exception. In his wonderful book *Thinking, Fast and Slow*, he notes that he has an optimal speed for thinking and working while walking.[12] He even suggests that the mild physical provocation involved in walking might generate greater levels of mental alertness. There is an interesting coda to Kahneman's account. Walk, and try working out 23 × 78 in your head. You will almost certainly come to a halt and stop walking. Kahneman comments that 'accelerating beyond my strolling speed completely changes the experience of walking, because the transition to a faster walk brings about a sharp deterioration in my ability to think coherently'.[13] Kahneman's example speaks, of course, to his own particular experience; the multiplication task is a well-defined and constrained rule-governed problem, the solving of which requires a strict sequence of subtasks, all depending heavily on maintaining simultaneous representations in working memory. In other words, a creative solution is not required – in fact, a creative solution would be profoundly undesirable. But is this the case for other types of problems? What about problems that don't yet have a definite answer, because the problems themselves are fuzzy, ill-posed or incomplete in their expression?

A counterpoint to Kahneman is the story of the great Irish mathematician Sir William Rowan Hamilton of Trinity College Dublin. Hamilton had grappled with devising a new mathematical theory – 'quaternions', which extend the mathematical theory of complex numbers to three-dimensional space. The mathematics involved are strange, and alien to everyday experience, but they find many contemporary uses in physics as well as in computer gaming, animation and graphics, and even in the design of electric toothbrushes.[14] Hamilton discovered the solution while on one of his regular two-hour walks from the Dunsink Observatory in north Dublin to Trinity College or to the Royal Irish Academy (both in the city centre). Hamilton inscribed the fundamental equation with his penknife on the stone of Broom Bridge in Cabra, Dublin. Today, a plaque commemorates his insight:

> Here as he walked by on the 16th of October 1843
> Sir William Rowan Hamilton
> in a flash of genius discovered
> the fundamental formula for
> quaternion multiplication
> $i^2 = j^2 = k^2 = ijk = -1$
> & cut it on a stone of this bridge.

Hamilton himself described this moment as follows: 'And here there dawned on me the notion that we must admit, in some sense, a fourth dimension of space for the purpose of calculating with triples ... An electric circuit seemed to close, and a spark flashed forth.'[15] Here are many of the essential elements of the moment of creative inspiration: the long period of contemplation, preparation and incubation, the active construction and formulation of a new problem, a testing of differing solutions through a prolonged period of thought. And, all the while, lots of walking. There is now an annual Hamilton walk on 16 October to celebrate the importance of his work, with mathematicians from all over the world participating.

What, then, are we to take from these anecdotes? Does walking present a barrier to or facilitate creative mathematical thought? They address different problems – one clearly defined with a single solution, the other ill-defined, requiring a creative solution. This suggests walking facilitates contemplating the latter problems for which divergent thinking is needed – thinking that requires exploring multiple potential solutions. Walking can affect the process of creative thinking during walking, after walking, or both. And it might do so in positive and negative ways, depending on when the effect of the intervention – the walking – is measured. There might be a long-lasting and oblique spillover effect on creative thinking, arising from the very fact that the brain is in a better physiological condition than it would be otherwise. Finally, we have to consider one more thing: as Kahneman acknowledges, walking at high speeds is effortful, and requires fairly continual attention in order to stop one breaking into a run, or merely to prevent stumbling and falling. Whilst we know already that being aerobically fit has multiple, profound, positive and enduring effects on the brain, just as it does on heart and lung function, it may therefore be the case that walking at a speed just below that which requires continual monitoring exerts the best possible effect on creative cognition.

Before we explore further the contributions of physical activity on creative cognition, let's ask what we really mean by the term 'creativity'? What is creative thinking? One widely accepted definition is that it involves two core concepts: the creation of something that is novel, and on which some value may be placed.[16] (These are subjective assessments, of course, and what may be seen as valuable at one time may not be at another.) Creativity is most often assessed by tests which focus on divergent thinking and convergent thinking, and occasionally on artistic expression. Divergent thinking typically involves generating a variety of solutions to a problem (a standard question is 'How many uses can you think of for a brick?') within a limited time period. Convergent thinking, by contrast, requires a single, unique solution that solves this particular problem exactly. A cryptic-crossword solution is an example of a convergent solution, as is Einstein's famous equation $E = mc^2$.

Whatever type of solution we need, how do novel or creative thoughts arise in the brain? A useful way of imagining the brain is as an enormously complicated network of cells, regions, and circuits, with varying degrees of traffic between the parts of the overall network, depending on the demands being made on the network. Areas that are remote from each other will tend to interact less, and areas that are close will interact more. Consider the extended network of people you call on to solve a range of problems, whether it's filing a tax return, or fixing a plumbing issue: you are likely to learn more from people that you don't interact with that much. The brain is similar when it comes to the process of creating new ideas. We need access to far-flung remote associations between different brain areas – for this is how new and interesting, creative ideas might arise.

It's clear that you must have plenty of knowledge regarding the problem – a well-stocked brain is an important precondition for creative problem-solving. Another, complementary, way is to get more of the brain active. One of the simplest means of doing this is to get up and walk about.

Standing causes immediate changes in both blood pressure and in brain activity. Standing and walking places greater demands on the body and brain than does sitting.[17] More oxygen is needed, and greater activity is required across many different brain systems to ensure that you don't fall over, so that you can see where you are going, and to coordinate your limbs. Then you must make quick microdecisions about the direction of travel – even if it is just a circumnavigation of your office. A simple, collateral effect of rising and moving is that activity spreads across more distant brain regions – increasing the likelihood that half-thoughts and quarter-ideas, sitting below consciousness, can come together in new combinations.

We can demonstrate that this theoretical scheme is correct by systematically testing novel ideas generated when sitting, standing and walking. Recent innovative experiments have shown that walking boosts creativity and problem-solving in a variety of unexpected ways. The psychologists Marily Oppezzo and Daniel Schwartz of Stanford University conducted a series of experiments in which participants completed tests of creativity.[18] The first was a divergent-uses test (for example the brick question), whilst the second was to create metaphors (describing an egg hatching as 'a budding cocoon', for example). Participants were either walking (on a treadmill or outdoors), seated, or pushed in a wheelchair. In all cases, walking of any sort substantially increased scores on both types of test, with the highest degree of novelty seen in those walking outdoors. These effects are robust: idea production increased several-fold for walkers, in a reliable and sustained way, compared with those seated. As the researchers concluded very simply: 'walking opens up the free flow of ideas'. These and similar studies suggest that walking is a powerful boost to creative cognition because of the particular way it entrains remote associations in the brain, in addition to the stimulus provided by nature.

Could walking affect creativity differently than aerobic exercise? Does walking offer a reliable method of generating creative thoughts that otherwise might not surface? We need to try and measure creative ideas during walking, and to measure creative ideas as a result of having walked. (While they may overlap, these are not the same thing.)

Some recent experiments have also sought to test whether mood, as well as movement, may play a role in raising creativity.[19] The question is, does elevating mood have an effect on creativity, independent of exercise; or does exercise have an effect on elevating creativity, independent of mood? To test this, participants were recruited to engage in either an aerobic workout or an aerobic dance, and to watch a rather unengaging video. Participants either watched the video first and then exercised, or vice versa. In both cases, mood was assessed using a checklist of positive and negative adjectives, and participants also completed a test designed to measure the fluency, originality and flexibility of their creative thinking. On average, a twenty-minute period of exercise was found to elevate self-rated mood by about 25%, whereas watching the video lowered their mood. They also found that exercise does have an effect on creativity, though the result was smaller than the exercise/mood correlation. Creative fluency and flexibility increased, though there wasn't an effect on originality. Overall, they concluded that physical exercise 'can be said to have slightly enhanced creative thinking ... independent of mood changes'.

One important consequence of the idea that we are cognitively mobile is recognising that our brains occupy space within our bodies: the inputs and outputs to and from our brains are mediated by our bodies. The fact that brain regions concerned with thinking, reasoning and imagination are cross-connected to brain regions concerned with planning and intended movement makes sense in terms of the neuroanatomy – and if you assume that the purpose of thought is action or movement. Hence the humble sea squirt needs a brain when it is motile, but not when it is sessile. But it is now becoming clearer that even our posture, as well as motion, might also have an effect on cognition, mood and creativity. It is hard to disentangle mood from this completely as common sense tells us that positive mood is associated with openness to experience, and that this in turn might be reflected in the form of the posture that we take. Nevertheless, a recent study asked participants to assume either a closed posture (arms and legs crossed tightly) or open posture (arms and legs spread out), and then to watch a positive or negatively valenced video.[20] It's important to remember that it is not the mood itself that is key, but the effect of the mood mediated by the posture adopted. When mood and posture are congruent, even a negative mood with a closed posture, creativity should be affected positively. And the effects noted by the study – which although not particularly strong were nonetheless present – suggest that creativity can be facilitated by adopting postures that are compatible with the mood that we are experiencing.

Studies of this type underline how profoundly what we are doing, and how we are doing it, affect creativity. They also suggest that the standard methods of assessing creativity used through the generations by psychologists and neuroscientists may be underestimating our capacity for creativity, because the environments that we test in, and the postures that we ask participants to adopt, constrain how they perform. This may equally be true of workers in their day-to-day environments. If we want to encourage freer forms of creative cognition, we need to get people up from their desks, away from their screens, and get them moving. This particular effect of movement on creativity is a powerful and largely untapped source of generating new ideas. Office buildings that provide indoor and outdoor spaces within which to roam and to converse must be encouraged, as should providing easy means of capturing thoughts on the fly. But the fundamental issue is legitimising, supporting and institutionalising this behaviour while at work. Expecting knowledge workers to generate deep and creative solutions to complex problems while seated in a crowded shared office is unreasonable and self-defeating.[21] Adopting strategies harnessing the power of mobile cognition will have measurable effects on a worker's mood and well-being, as well as on their productivity. There is no reason not to try, apart from behavioural and organisational inertia – which is, of course, the hardest thing of all to overcome.

*

We can reach a more creative state, we now know, by being in motion. And while that creative state can help us solve different types of problems more inventively, what we also find is that it can shape our experience of the world – it can, for example, change the nature of how time is experienced.

If you are driving through countryside on an open road, a three-hour journey may feel like it passes in no time at all. On the other hand, a three-hour journey on a jammed motorway will almost certainly feel much longer. Our perception of time passing does not have the consistency of a clock – our psychological units of time are not the same as chronological units of time. In this case, pleasure affects perception. Certainly, a walk undertaken at the height of the noonday sun while you are thirsty, tired and hungry will feel more onerous and longer than a walk in cool conditions when you are rested, fed and hydrated. Humans commonly underestimate or overestimate time intervals depending on how we are feeling. Albert Einstein captured this well: 'Put your hand on a hot stove for a minute, and it seems like an hour. Sit with a pretty girl for an hour, and it seems like a minute. That's relativity.'

But, does our estimation of how we feel time passing vary when we are walking compared to, say, sitting? To answer this, we need to turn to that old workhorse of studies of human walking, the treadmill, and to use it while participants engage in time-estimation tasks. In a time-estimation task, you are asked to push a button after you believe a certain passage of time has occurred. This might be an interval as short as one second, it might be five seconds, or a minute, or longer. By varying what we are getting participants to do while making multiple time estimations, we can see if the task changes their estimates. We can also test if there are reliable differences in time estimation between individuals.

In one study, participants were walking (a motor performance task), and performing a cognitive task, namely making estimates of one-second intervals.[22] A metronome set to a one-second pace was played to participants for the first thirty seconds of the trial. The participants were instructed to press a button in sync with the metronome as practice in timing estimation while either seated or walking on the treadmill. If walking is principally governed and controlled by the central pattern generators in the spinal cord, then there should be relatively little effect of time estimation on walking speed. This is because our inner metronome – and not the movement of our legs – is setting the beat. And this inner metronome can work in several ways: a top-down instruction from the brain; pacing by the central pattern generators in the spinal cord; and walking itself, or any feedback that arises from walking. Time-estimation tasks here will allow you to test these possibilities. The results confirmed that time estimation varied with walking speeds. Thus, walking itself changes the estimates of subjective units of time that the participant experiences – our experience of time varies depending on whether we are walking or sitting.

Our experience of time can vary widely between two poles. The experience of time passing ever so slowly is familiar to the walker tramping slowly and arduously uphill in the rain with a heavy back-pack. In contrast, when the temperatures are moderate and the trail is downward-sloping, then occasionally the experience is such that huge areas of ground are covered for what feels like minimal effort and great enjoyment in comparatively short periods of time. This state is known as 'flow', and it can apply not only to walking but to a huge range of activities, to work, to sport, to specialised performance of all varieties. Sometimes referred to as the psychology of optimal experience, flow is a central psychological concept, first developed by Mihaly Csikszentmihalyi.[23] Flow is the subjective experience of concentration and deep enjoyment accompanying or arising from skilled performance. Feelings of control, of oneness, of immersion, of being in the zone, are all characteristic of flow. The experience here is very different to the elation of overcoming difficult terrain and sitting down for a well-earned rest; rather, it is in the freeing of the mind and brain from the moment-to-moment control of walking, while at the same time covering considerable distance.

We humans are skilled, expert, accomplished walkers: walking is a prime means to experience flow, available to almost all of us. Walking can facilitate the flickering between differing states of mind, and it is this that makes creative cognition most possible, precisely because walking allows a peculiarly enjoyable *mindlessness* (not *mindfulness*). Walking, with the experience of no particular thought to engage you, is one of those opportunities when odd but creative associations might arise across differing parts of the brain's semantic networks that process memory and meaning.

There has been an unusual and interesting study of walking and thinking among a small group of Norwegian academics.[24] The participants were chosen because they all like to undertake extended walks, and to do so regularly. Moreover, they identified walking as an important aid to their thinking. It is difficult to generalise widely from such a small sample, but nonetheless, some instructive points emerge from the semi-structured interviews that were conducted. All of the interviewees thought that walking at a certain pace and rhythm was most conducive to thinking. All interviewees emphasised the word 'rhythm' in different ways. The optimal speed which facilitates thinking varied, but by common consensus among the group it was 'a speed when your body is engaged and stimulated, not overly taxed'. Another common theme was that they thought of thinking as being almost like a 'place'. One interviewee said:

> I feel like I am in it, in this kind of weird abstract world of the things I have read, the things I have thought while reading. I think of it as a state, but I also think about it geographically, that you are in a place where all this knowledge surrounds you in these weird more or less solid shapes, so you can sort of summon this book and this argument and this thought, and this writing that you linked to that author, and it is all available to you. You are in the middle of it and you can call on thoughts.

What we see here, as we do throughout this book, is that walking brings us to a place of clearer thinking. We can think of ourselves as being able to walk away from a problem by getting to a place where a solution is possible. It is a peculiar and wonderful creative problem-solving state comparable in many ways to that which can arise on the edge of dreaming, and even during dreaming itself. The phrase 'sleep on it' is an everyday testament to sleep's generative and creative powers, and writers through the ages have attested to sleep's great problem-solving properties. John Steinbeck wrote that 'It is a common experience that a problem difficult at night is resolved in the morning after the committee of sleep has worked on it.'[25] Scientists have found the same: one famous example is August Kekulé having a dream of a snake eating its own tail, and realising that the molecular structure of benzene must be a ring-shaped.[26] Beyond anecdote, though, experimental studies support this connection. In one study participants were given a difficult cognitive task requiring the learning of stimulus-response sequences. Hidden within the task was a shortcut – an abstract rule that allowed quick solutions. Exposure to the problem followed by eight hours of sleep more than doubled the probability that participants would solve the task, underlining how emphatically sleep provides time for offline problem-solving and memory consolidation.[27]

The challenge with dreaming is of course that the thoughts can be too fleeting to be captured, but the characteristics of dreaming, the loss of meaning of time, the experience of reverie, the free association involved between differing memories and thoughts, might still, in part, be created during walking. The regularity of the rhythm of walking itself, paced by the spinal cord's pattern generators, coupled with a de-emphasis on time, and time itself, perhaps provide a good way to kick-start the kind of creative thinking we all need. So, the next time you have a difficult problem to solve, tell your boss you won't be at your desk for who knows how long (after all, your time estimation will be affected); you're going for a walk, and when you come back, the problem might have been solved. They may not be convinced, but urge them to try it themselves. Be sure to take some way of recording your thoughts. Bring some notes, scratched on a page. Perhaps go with someone who also cares about solving the problem – and chat in a focusedly unfocused, recurrent way, taking breaks to talk about other issues. You must, of course, have a well-stocked mind, with lots of information about the problem that you are going to solve. $E = mc^2$, or $i^2 = j^2 = k^2 = ijk = -1$, will not come unbidden to you, for these solutions require an intense period of preparation, reflection and focus. You should walk with the purpose of thinking about the problem. It doesn't matter if you don't solve it immediately. Walk without expectation of a solution. Instead, walk for the enjoyment of walking, and for the enjoyment of thinking about the problem.

There is a powerful lesson to be learned here: those charged with complex political, organisational and other problems should not be cooped up in conference rooms. They should get out and walk their way to better solutions, and to a better world.

# 8. SOCIAL WALKING

There are some walks I have avoided. I have never walked a pilgrimage; I almost certainly never will. To my knowledge, I have never walked in my sleep – and hope never to. I have never walked the plank, or the red carpet. I hope to avoid both of these also. Nor have I attended many protest marches.

Many of the very best walks of my life have been with friends and family, usually in cities, but often in the countryside too. A number of my best memories of walking are doing so on cold, sunny, days: a little cold helps to avoid overheating, and a little sun brings everything into sharp relief. And the conversation flows easily. I especially enjoy walking at night. Walking at night might seem a bit odd, a little asocial perhaps, but it offers something special too, especially when you're out walking with a friend through the quiet streets stripped of traffic, glimpsing brief snatches of life and light while walking past windows, looking up to the stars, knowing a comfortable bed awaits. There is something otherworldly about nightwalking, something that makes you see the world differently. Thomas

Kinsella, the great poet of the urban, wrote in his epic poem *Nightwalker* that at night-time, when walking 'the shadows are alive. They scuttle and flicker / Across the surface'. On your own this might be frightening, but for me, especially when I have company, it's a transformation that is wondrous: the mundane and quotidian are rendered anew.

What binds these walks together is something often unrecognised and unacknowledged: at its heart, walking has a profound social function. We walk together for an ideal; we walk together to source food we will share; we walk together for social display; we walk together to try and change the world; we walk together to find better lives for ourselves and each other; we walk together to enjoy each other's company. We have evolved to walk together, and social walking is demonstrative: it sends signals to others about our shared intentions and collective goals. Social walking can be the best of walking, whether towards a common goal, or just sauntering along with no particular place to go.

Yet we so often overlook the extent to which walking occurs in groups – it might be a nuclear family, an extended family, a crowd of teenagers in hoodies, old men pulling golf clubs around, or ranks of soldiers walking in step. It is a great and profound mistake to think of walking as a purely solitary activity. As we've seen earlier in this book from our ancestor Walking Eve, our relationship with the act of walking is one that stretches back into deep time. We see traces of this social inheritance in the fossilised footprints of a walking group of some eighteen people dating from about 19,000 years ago – early modern humans.[1] A group are walking together, across a mudflat near the base of a volcano, close to what is now called Lake Laetoli in Tanzania. The pattern of footprints, scattered across an area roughly the size of a tennis court, is complicated, but has yielded some of its secrets. The group comprises mostly females and children. Their feet are sinking into the mud. The water falls from their feet, leaving traces of drops between steps. One of their number has what seems to be a broken toe.

They are an extended, interacting, social group, presumably carrying some food and water, and perhaps elementary weapons for protection. Possibly they were clothed, but this is all we know of them as individuals. And but for a thin, silvery volcanic discharge which covered their prints, they would have been completely lost to deep time, their lives, their flesh, their bones. They have left an echo, a trace of what they were. Ancient evidence from other times and places shows similar social walking – in Australia, England, Argentina, Nicaragua. Without these larger walking groups, there is no reproduction, no specialisation, there are no coordinated attacks on other groups to steal food or other resources, no colonisation of new places, and no new horizons.

There are also of course some walks which are conducted alone but which have a profoundly social dimension. Pilgrimages are a notable example of social walking, even when they are performed by the solitary pilgrim. They are performed in solidarity with a greater purpose – the community, a cause, a faith. They demonstrate the power of an ideal to animate a walking tribute to it. Even the solitary pilgrim is walking for, and with, an imagined community of the mind. There are the solitary figures of the *boulevardier* or the *flâneur* who find purpose in the social fabric of the city, observing the crowds and spending time dipping into public spaces and crowded places.

Social walking manifests itself in many other positive and powerful ways, and is crucial to creating or maintaining social cohesion at an intimate one-to-one scale, and for wider society. Humans commonly hike on trails together, or wander the city with apparent aimlessness while chatting, or go on protest marches together. Walking together offers a chance for conversation to evolve in ways that it couldn't, indeed wouldn't, if you simply sit together. Mark Twain expressed this idea elegantly:

> The true charm of pedestrianism does not lie in the walking, or in the scenery, but in the talking. The walking is good to time the movement of the tongue by, and to keep the blood and the brain stirred up and active; the scenery and the woodsy smells are good to bear in upon a man an unconscious and unobtrusive charm and solace to eye and soul and sense; but the supreme pleasure comes from the talk.[2]

Walking can be central to our sense of connection to other people and the world around us. One recent major study of the elderly concluded that those who spend approximately 150 minutes walking per week are more socially active and have a sense of better overall well-being than those who are less active.[3] This correlation between social activity and general well-being is one that has been shown in a huge range of studies. A low-tech innovation for public health policy could use text alerts or social media groups to create regular walking groups for the elderly – the benefits would be great, and the costs negligible.

This social dimension of walking is evident from an early age too. Learning to walk changes forever the quality of our social interactions. As crawling babies, our range of head movements is restricted; down on all fours, your view is of the ground, and to see your caregiver clearly you need to sit in order to look up. On two legs, things change: you can see each other easily, without awkward postural changes. Learning to walk fundamentally changes the type of social interactions and gestures that children can engage in.[4]

Studies that have examined walking children, crawlers and children in baby walkers, and compared their interactions with caregivers and toys, have shown that walkers play with their toys more, and they vocalise much, much more. Children learn to use gesture, sound and movement to engage a caregiver socially in play. Freeing the hands is liberating for a child, and it allows them to free their minds. Once we become walkers, our opportunities for interacting with each other, for any transaction, even simple food sharing, increase dramatically. This is also the case, as we've seen in earlier chapters, when we compare going around on foot to travelling in a car. On foot we are capable of interacting with each other at a human level: we quite literally have more common ground, we can synchronise more easily, and we can have shared experiences, including the same environmental conditions like the weather (a subject known for bringing strangers together).

More than that, as we have seen, walking affords another possibility, which is that for collaborative creativity and also for playfulness. Think about taking children out for a walk. They are highly active walkers, sometimes riotous and unconstrained, and rightly so. Their example should be a spur to us all to enjoy walking – to walk with enjoyment, and not simply as a means to get from one enclosure to another.

*

Social walking is a curious phenomenon: the astonishing feat of mobile brain and body coordination it involves is often overlooked. Social walking requires that we fall into step with one another, allowing us to maintain a common behavioural purpose for some period of time. It involves coordinated and simultaneous action in multiple brain regions to control one's own trajectory and direction of movement, and predict the trajectory and direction of movement of those who you are walking with. Critically, each individual must use these predictions to try and simultaneously synchronise their movement with that of the other person or group while often doing something else – like talking or singing or chanting. This is a difficult problem – so much so that robots can't yet do it! However our brains can solve this problem quickly and easily – most of the time.

We are acutely sensitive to social signals provided by other people: this sensitivity depends critically on the rapid, sub-second action of two brain systems – the 'mentalising' network[5] and the 'mirror neuron' network.[6] The mentalising network is judgement-orientated – it allows you to draw inferences about 'agency' in others, in other words the extent to which you judge they will follow through on their intentions. The mirror-neuron network is action-oriented: it responds to your own movements, and similar movements by another in the same way – it signals the movements of others. The neurons of the mirror system respond, for example, when you extend your arm forward to shake the hand of another person – and it also responds to another person reaching to shake your hand. Putting these systems together allows the prediction of the direction of movement or trajectory of the other person while you walk with them. To engage in social cognition requires the unconscious exchange of signals such as speech, expressions, posture and body movements, and it creates a shared world through joint attention and interpersonal synchronisation supported by brain systems that represent the agency and actions of others.

So we are able to walk side by side – but why do we want to? Why does this synchronisation generate a feeling of connectedness? In short, why does social walking feel good? Interpersonal synchronisation is the name given by psychologists to this phenomenon. Interpersonal, because it involves two or more people; synchronisation, because naturally and unconsciously we start to mimic each other's gait, which entrains other, deeper processes in the brain and body: our breathing becomes synchronised, our heart rates must perform similar functions at similar times, and our brains simultaneously take account of what it is that the other person is likely to do, as well as monitoring and controlling what it is that you yourself are doing.

We find it relatively easy to sync with one or two other people, but walking in a larger group requires more of a conscious effort at interpersonal synchronisation, sometimes helped by the natural or deliberate emergence of a leader or a lead group. A natural dynamic, which expresses itself most commonly in conversation, usually occurs here. When a group of three people speak, each can pay attention to the other. If a fourth person joins them, after some moments usually the group will divide into two pairs. If the group consists of five individuals, it will typically break into a group of two and a group of three – this seems to be the maximum number that we can pay attention to effortlessly and easily during conversation. And the same seems to be true in groups of walkers.

This leads us to a problem: our natural inclination will be to divide into groups of two and three people, but marches or other forms of large and organised walking groups must be coordinated, so that pace, direction and unity are maintained.[7] One easy route to group coordination is through the use of our voices. Singing or chanting in unison while walking together in large groups has been practised since time immemorial. Singing and chanting might help each of us coordinate our walking with each other, providing us with a kind of a voice-originating metronome for timing our walking. If you observe marching groups or people engaged in demonstrations, they will often be accompanied by chants, drum beats, or some other type of sound to facilitate synchronisation.

Language, according to the cognitive primatologist Robin Dunbar, is a key element in the formation and maintenance of human social groups, rather like the function of grooming in our non-human primate cousins.[8] The signalling properties of spoken or sung rhythmic language, enjoining us to listen up, to look here, to move there, make it the perfect medium for coordinating the behaviour of larger groups. If we all want to participate in the conversation, we must remain in earshot, and we must speak sufficiently loudly so that everyone can hear. This causes the group to huddle, and it will cause the members of the group to maintain pace with each other.

This coordination problem, of course, breaks down once the group starts to get very much larger. But a regular rhythm can still mitigate the size problem. If an external pacemaker is provided, synchronisation should be possible across differing types of behaviour, be it walking, clapping, or singing. In the case of a sound stimulus used to coordinate individual walkers to move in unison, we would expect to see activity not only in the parts of the brain concerned with hearing (the auditory cortices) but also in those parts of the brain concerned with planning and executing the movement itself (the motor system, including motor, premotor and supplementary motor cortical areas).

The timescales of coordination matter: when we clap in unison, for instance, we do not do so with millisecond precision (because we all differ a little in our hearing, speed of movement and other capacities). Interestingly, an experiment that asked groups of people to synchronise finger tapping revealed that participants who displayed greater levels of social anxiety also demonstrated a decreased ability to synchronise with the group.[9] So anxiety about how you will perform a simple social synchronisation task militates *against* performing that task. Even on this tiny, fingertip scale, stage fright is a real phenomenon, causing people to choke. Synchronising finger tapping is not explicitly social in the sense of one person conversing with another. Instead, the task is implicitly social: your task is to behave as the other person behaves. This social dimension to the activity was evident during the study from significant increases in activity in parts of the brain comprising the brain's 'social network', as well as in brain regions concerned with introspection and self-reflection.

This study was on the micro scale – it involved small groups of people making tiny movements with their fingers. Imagine, for a moment, being part of a large, dense crowd, perhaps at a station or sports event. One of the regularly encountered difficulties in mass public transport systems, such as airports, rail systems, or other locations, is ensuring that people walk at a steady rate, coordinate their pace with the person in front of them. Ideally, pedestrians on such systems should all walk in the same direction, ensuring walking flow is not mixed. How might you ensure that walking flow is steady and stable, indeed predictable, through time?

Another revealing study explored how the brain engages in social walking on this larger scale.[10] This experiment used technology known as near-infrared spectroscopy (NIRS), which allows a gross measure of oxygen uptake in certain brain regions, the theory being that brain regions which are more active will probably consume more oxygen. Participants wore a head-mounted NIRS device while performing group coordination tasks during walking. Some groups walked with the sound of a metronome set at seventy beats per minute (approximately the resting heart rate of a young, healthy adult). The other group weren't given any sound beat at all. The first group were noticeably better at coordinating their walking, and at managing to maintain a steady pace. There was also a substantial increase in activity in the frontal lobes of the paced-sound group, compared to the group without the paced sound. The frontal lobes are generally involved in formulating intentions and planning actions, suggesting that a regular beat aids synchronisation of the networks of the frontal lobes that may be involved in planning and goal selection.[11]

These kinds of studies suggest that providing a sound source pulsing at about the average human heart rate might ensure a steady flow of walking in large groups of walkers. Slower or faster rates may mean people change pace accordingly. This means that providing unobtrusive but audible sounds in synchrony with the normal human heart rate might help manage pedestrian flow in busy transit systems, at concerts, or even in an evacuation.

How might sound pace the activity of walkers? Studies have shown that simply watching others engage in movement activates the motor system in the brain; for example when watching a foot being moved up and down, the area of motor cortex that controls foot movement was activated in the observer.[12] Mimicry of the behaviour of others, and preparedness to mimic others, is pre-built into the nervous system. We automatically and unconsciously prepare for socially driven synchronisation of motor movements, be they body parts, or the whole body. (This preparation does not intrude into consciousness, unless we deliberately take a top-down decision that we are going to mimic the behaviour of another.)

These studies confirm what we already know intuitively from our everyday lives: humans are exquisitely sensitive to the behaviours of others, and engage in rapid interpersonal behavioural synchronisation. These capacities are also demonstrated from an early age. Very young infants are able to track the eye movements, head movements and hand movements of significant others. Moreover, they are able to mimic the gestures that others make to them. What happens where walking is concerned? At what age do children become sensitive to the walking trajectories of others? It seems children as young as four are as sensitive to the direction of walking as adults.[13] This was revealed by an experiment that placed four-year-old children in front of computer screen that showed a simple walking figure, who would walk from the top to the bottom of the screen. If their path was uncorrected by the child, the figure would walk into a tree or a house. The onscreen walkers were all in two dimensions, as were the objects at the boundary. Therefore, there were no three-dimensional depth cues available to help the children estimate the likely walking trajectory.

The estimation of walking trajectory therefore had to be made on the basis of the movement information contained in the two-dimensional walking figure alone. Amazingly, children performed in the same range as adults and could discriminate very tiny trajectory differences.

Sensitivity to the walking direction of others is a profoundly important social cue, one vital to navigating crowds without colliding into each other. Sensitivity to the walking direction of other walkers is required for intersecting with those walkers (for example, as might occur during tackling in a ball game). That these sensitivities to walking in others appear so early in development suggests again that, at its core, our walking has a profoundly social function, as in when the child runs to and is swept up in the arms of a caregiver, dodges a beast of prey, runs to hug another child or tackle them in a game. Our walking serves much more than individual mobility – it also serves a profound source of social interaction with other human beings.

*

As noted earlier, one of the great but overlooked lessons of our walking lives is that walking evolved with a distinct social purpose, such as migration and exploration. We also walk together in protest – at the decisions of political regimes, or against an individual or in response to a violation of rights – which is something fundamental to our society. This willingness to walk together, to give collective expression to our like or dislike for something, is a truly central feature of being human, and is not something shared with even our closest relatives. From where does this willingness to walk and march together arise? Research has shown that we can experience a psychological high from being a unit of a large crowd assembled for a common purpose – whether for a protest march, a concert, a religious ritual or a sports match.[14] People who are able to feel part of a crowd report at least transient increases in their feeling of well-being.

This feeling of 'effervescence' from collective activity has been measured through a self-report questionnaire, comprising a series of simple statements such as 'I feel connected to others when in a large group activity I like, like going to a concert, church, or a convention', or 'When I attend a wedding I feel a connection to the other people there.' Respondents gave a response to each question on a seven-point scale, from strongly disagree to strongly agree. Females scored slightly higher on the scale than males, and people who self-rated as more religious scored higher than the non-religious. When assessed against other self-report measures, people who scored highly on the questionnaire had lower levels of self-reported loneliness, higher levels of positive feelings, higher levels of meaning in their lives, greater levels of self-awareness, and finally were more likely to feel collectively and relationally interdependent. This 'effervescent assembly', or the psychological benefits of group activity, is a real social phenomenon, and it correlates with other aspects of life that are involved in social connections. Overall, it transpired that life outcomes appeared to be better in those individuals who experienced high levels of social connectedness while walking.

Because the experience of being in a marching group can be so emotionally intense, we can come to believe that assembly and marching changes the societies we live in – which is not always the case. In free societies, we are able to participate in large-scale assembly, but the downside is in mistaking the feeling of unity, interdependence and relational outcomes as actually influencing political processes or policy outcomes. The hard realities of legislation or policy are often unaffected. Sometimes, though, marches can evolve into something else: mass demonstrations before which autocrats find they are powerless, because assent is completely withdrawn from them by both the people and the security apparatus of the state. The mass demonstrations and marches that preceded (and perhaps even caused) the collapse of the Communist bloc in 1989 are a profound example. Marching is not pointless if it is coupled to other forms of effective collective action focused on changing laws and policies.

The mass demonstrations organised by Gandhi demonstrated to all that removing consent by the governed makes colonial power untenable. The civil rights marches of the 1960s in the United States understood this lesson well. Mass marches, designed to demonstrate the depth and breadth of feeling on the one hand, but coupled to legislative action and the civil rights acts on the other, led to deep and enduring changes in the treatment of minorities in the United States. Civil rights marchers also undertook demonstrations across Northern Ireland in the late 1960s in what was seen as open defiance of the then Stormont Parliament. During one major march, the protestors were attacked at Burntollet Bridge in 1969 in what was characterised by one historian as 'the spark that lit the prairie fire' in Northern Ireland.[15] The 'prairie fire' was a three-decades-long conflict which consumed nearly 4,000 lives, and the scars from which have yet to heal. If the march had been allowed to pass, then the history of Northern Ireland might have been different, but as this is a historical counterfactual, we will never know.

Of course, there is an interesting lesson here: authorities should allow marches to pass off freely. Don't police them, except to prevent damage to life, limb and property. Unless they are designed purposely to bring civil society regularly and completely to a standstill, marches may act as a vent which releases the energy to generate social and political change. Autocrats, of course, despise free marches and free assembly, and will suppress marching with firepower, and have frequently done so. It's curious, of course, that the one form of collective walking that autocrats do approve of, the marching of soldiers, is a show of martial strength, with all individuality excised. Free assembly, and the free walking implied, gives life to the idea that the power held by others over us is because we assent to that power. Marching can give life to solving the collective action problem – how do we know that we all think and feel the same way about a crucial issue? Well, by getting out there, walking in unison and proving it on the streets.

As we've seen throughout this book, walking itself is an activity that is key to so many aspects of individual lives and societies. Walking should be central to policymakers, medical professionals and town planners. Walkers everywhere need a charter that is at the foundation of our communities. At its core this charter needs some straightforward principles and to have its provisions legally implemented. Earlier I suggested the acronym EASE as a tool for our town planners and architects (walking in our towns and cities should be *easy, accessible, safe* and *enjoyable*). Design principles built around EASE will increase the quality of all our lives. This is something we can both aspire to and convert into actual policy if politicians know that your vote depends on their support for a walker's charter. The principles of this charter need to be translated into 'on the ground' designs – not as an afterthought – but as the central design component.

The message should be straightforward: walking is good for you. But we humans are also a little bit lazy and resistant to messages which ask us to re-evaluate our world views. Deep policy documents are of use when attempting to argue or bolster a case.[16] To ensure change, the best methods are evidence-based appeals to reality and truth, within the context of a succinct story that adds to or coheres with values that the listener already has.

However, if mere information provision was enough, there would be no anti-vaccination campaigns, there would be no smokers, and we would have no problem with obesity. A walking campaign must have a few simple, straightforward take-homes in terms of behavioural change, such as ensuring the permeability and walkability of our towns and cities, providing well-designed green spaces for walking, promoting the needs of walkers as the central human experience of mobility and movement. We need to ensure our towns and cities allow us to do what comes naturally to us: afford us opportunities for lots of activity, as well as places to rest and recharge. Creating, engineering and defending walking places and spaces is the present and future challenge. Rising to it will enrich us all in more ways than we know.

# AFTERWORD

In the course of this book we have travelled back into the depths of evolutionary time to meet our bipedal ancestors. We've witnessed how walking is a wonderful solution hit upon time and again by nature – from bottom-dwellers on the ocean floor and undulating tetrapods seeking food on sandy beaches, to we humans seeking new worlds to conquer, undertaking those great migratory waves out of Africa to walk the world over. We've learned that to avoid getting lost we need cognitive maps that work best when they are regularly activated by walking.

We've wandered about the marvels that are our modern cities. Our cities can be the best of walking – if they are designed with walking in mind. We need to conceive of walking in cities in its broadest sense – giving EASE for the elderly, the young, those who have to use walking sticks, crutches and wheelchairs. As we become an increasingly urban-dwelling species we need to remember this – our cities are for people.

We've journeyed back to childhood, when we learned to stand, struggled to our feet, fell down, and struggled back up again. And we've learned how all of our senses are sharpened by walking, freeing our hands for gesture, for tool use, for carrying food and children, our feet rhythmically moving, swinging through the air, stabilising against the ground, gaining purchase, and onwards we move again. The commonplace wonder is that we do this largely automatically.

How can you make these lessons endure beyond reading this book? For one thing, use a walking app. Turn on the alerts. How many steps have you walked today? How many last week? Not only can you measure your own walking, you can compare yours to that of your friends, your age group, the national average. And all you need do to walk a little more is park your car a bit further away; get off the bus one stop earlier; walk to the shops; walk to work; walk to school.

We know that walking improves your mood, and more than you think it does. Walking might also be a kind of a behavioural inoculant against depression, as well as against the slow, malign changes that mould your personality for the worse because you are sedentary. And walking also brings with it marvellous problem-solving powers. Your creative impulses, fostered by walking, will help knock the problems of life over.

The core lesson of this book is this: walking enhances every aspect of our social, psychological and neural functioning. It is the simple, life-enhancing, health-building prescription we all need, one that we should take in regular doses, large and small, at a good pace, day in, day out, in nature and in our towns and cities. We need to make walking a natural, habitual part of our everyday lives. Pound the pavements; get the wind on your face; let the light of day and street lamps of night dance on your eyes; feel the rain on your face; sense the ground beneath your feet; hear the sounds; talk – if only to yourself; relax into the rhythm of walking and let your mind wander, deliberate, contemplate; journey into your past, delve into your possible futures; or think of nothing at all. Although walking arises from our deep, evolutionary past, it is our future too: for walking will do you all the good that you now know it does.

# ACKNOWLEDGEMENTS

This book arose directly out of a great and revealing conversation I had with my ever-wise and learned literary agent, Bill Hamilton of AM Heath, in October 2016 in the Wellcome Collection in London. I hadn't realised I should write this book until that revealing, interesting, important conversation. For that, and much subsequent guidance – thank you, Bill! Thank you also to the team at AM Heath, especially Jennifer Custer and Hélène Ferey, who have done so much work as well to help this book see the light of day. A major thank you, too, to Stuart Williams of The Bodley Head, who saw the value in this book, and gave it his support to see it through. I could not have had a better editor and publisher. Thanks too to Anna-Sophia Watts and Lauren Howard, both of The Bodley Head for their close readings of the MS.

My two previous books have dealt with quite differing topics – but there is an underlying unifying theme to them and this book also: that of the 'brain in the world', taking a 'brain's eye' view of the world. One of my research concerns is applying psychology and neuroscience in policymaking and related areas. It is the most wonderful time to be interested in the science of brain and behaviour because of the wonderful being undertaken worldwide. My hope with this book is that concerned policy makers – in organisations, government, business, or wherever – take science and evidence more seriously in deciding and implementing public policy. And to do what scientists do – conduct experiments, test ideas, and let bad ideas die when the evidence shows they are wanting. The ideas in this book are based on my reading of the peer-reviewed literature: while they might be controversial for politicians, architects, urban planners and road engineers, they are not controversial in psychology and neuroscience. We need urban planners and engineers to embrace walkability as the core activity that our cities and towns revolve around and depend upon – for all our sakes.

I have enjoyed walking with so many people over the years that it would be invidious, indeed impossible, to name them all. A few though, I should. My parents, Mary and Rory, for being there through the many falls – of which I have no memory – that ensured I ended up being able to walk; Maura, my wife, for so many enjoyable walks in so many places over the years (but I'll mention in particular our many walks along the Promenade, in Salthill, along the edge of Galway Bay), now part of the Wild Atlantic Way; Radhi, our daughter, who always forgets how much she loves a walk until she is actually walking. All the Donnellys for so many great walks around Westport; I should also mention Myles Staunton for his pioneering work on establishing the Westport Greenway. John Miller and the late and much-missed Vincent McLoughlin, both of whom took me for my first and many subsequent walks through the Wicklow Mountains. Michael Gilchrist for many walks in Wicklow and Waterford. Vincent Walsh for many wonderful long and winding walks through the greatness of London; many more in the future, too, I hope. Ted Lynch for assistance with Latin and with whom I did so many walks around the wonders of Paris. Enda Kearns for the many regular evening walks around beautiful Killiney Hill and Dalkey. And to so many others, over the years, in so many other cities.

Thanks to the readers of the draft manuscript of this book, who improved it and challenged me as a writer: Jennifer Rouine, John Miller, Vincent Walsh, Robert McKenna, Charlotte Callaghan, Fiona Newell, Giovanni Frazzetto, Ted Lynch, Bill Hamilton and Jennifer Custer.

Susan Cantwell provided wonderful secretarial assistance, for which I am profoundly grateful, and which greatly speeded the delivery of this book.

I also thank the Wellcome Trust and Science Foundation Ireland for their generous support of my research through the years. Trinity College Dublin also deserves particular thanks for being a wonderful institution and a great place to work. As usual, any mistakes in the text are mine alone, and I apologise for them in advance. I have, of course, had to make decisions regarding inclusion and exclusion of certain topics from the scientific literature, and similarly regarding the selection and range of themes explored. Unfortunately, the writing has to stop somewhere, and some things inevitably get left out (such as the recent astounding advances in spinal-cord repair, which are worthy of a book of their own[1]).

# NOTES

## INTRODUCTION

**1.** Fitch, W. T. (2000), 'The evolution of speech: a comparative review', *Trends in Cognitive Sciences*, 4(7), 258–67, http://citeseerx.ist.psu.edu/viewdoc/download?doi=10.1.1.22.3754&rep=rep1&type=pdf.

**2.** There is a vast literature on the evolution of human bipedalism. The following is a small sampling. Thorpe et al. (2007), 'Origin of human bipedalism as an adaptation for locomotion on flexible branches', *Science*, 316(5829), 1328–31, http://science.sciencemag.org/content/316/5829/1328.long; Sockol et al. (2007), 'Chimpanzee locomotor energetics and the origin of human bipedalism', *Proceedings of the National Academy of Sciences*, 104(30), 12265–9, http://www.pnas.org/content/pnas/104/30/12265.full.pdf; Schmitt, D. (2003), 'Insights into the evolution of human bipedalism from experimental studies of humans and other primates', *Journal of Experimental Biology*, 206(9), 1437–48, http://jeb.biologists.org/content/jexbio/206/9/1437.full.pdf.

**3.** Roboticists are trying, though. See 'Robot Masters Human Balancing Act', https://news.utexas.edu/2018/10/02/robot-masters-human-balancing-act for a promising start, and the Boston Dynamics Big Dog is a remarkable robotic quadruped (https://www.bostondynamics.com/bigdog).

**4.** Straus (1952), 'The Upright Posture', *Psychiatric Quarterly*, 26, 529–61, https://link.springer.com/article/10.1007%2FBF01568490.

**5.** Richmond et al., (2001) 'Origin of human bipedalism: the knuckle-walking hypothesis revisited', *American Journal of Physical Anthropology*, 116(S33), 70–105, https://onlinelibrary.wiley.com/doi/pdf/10.1002/ajpa.10019.

**6.** Abourachid and Höfling (2012), 'The legs: a key to bird evolutionary success', *Journal of Ornithology*, 153(1), 193–8, https://link.springer.com/article/10.1007/s10336-012-0856-9.

**7.** https://www.bbc.com/news/uk-scotland-north-east-orkney-shetland-45758016.

## 1: WHY WALKING IS GOOD FOR YOU

**1.** Woon et al. (2013), 'CT morphology and morphometry of the normal adult coccyx', *European Spine Journal*, 22, 863–70, https://link.springer.com/article/10.1007/s00586-012-2595-2.

**2.** Rousseau, J. J. and Cohen, J. M. (1953), *The Confessions of Jean-Jacques Rousseau* (Penguin).

**3.** The psychologist Martin Conway argues that 'for many experiences simply recalling the meaning or the gist may be sufficient'; https://old-homepages.abdn.ac.uk/k.allan/pages/dept/webfiles/4thyear/conway%202005%20jml.pdf.

**4.** Stroop, J. R. (1935), 'Studies of interference in serial verbal reactions', *Journal of Experimental Psychology*, 18(6), 643–62, doi:10.1037/h0054651.

**5.** Rosenbaum et al. (2017), 'Stand by Your Stroop: Standing Up Enhances Selective Attention and Cognitive Control', *Psychological Science*, 28(12), 1864–7, http://journals.sagepub.com/doi/pdf/10.1177/0956797617721270.

**6.** Carter et al. (2018), 'Regular walking breaks prevent the decline in cerebral blood flow associated with prolonged sitting', https://www.physiology.org/doi/full/10.1152/japplphysiol.00310.2018; Climie et al. (2018), 'Simple intermittent resistance activity mitigates the detrimental effect of prolonged unbroken sitting on arterial function in overweight and obese adults', https://www.physiology.org/doi/full/10.1152/japplphysiol.00544.2018#.XCgVpEcW2lg.twitter.

**7.** Horner et al. (2015), 'Acute exercise and gastric emptying: a meta-analysis and implications for appetite control', *Sports Medicine*, 45(5), 659–78; Keeling et al. (1990), 'Orocecal transit during mild exercise in women', *Journal of Applied Physiology*, 68(4), 1350–3.

**8.** The hippocampal formation shows quite remarkable plasticity in response to aerobic exercise. A variety of data from differing groups shows that this effect is reliably induced by regular aerobic exercise: interventions promoting heart health also promote brain health; Erickson et al. (2011), 'Exercise training increases size of hippocampus and improves memory', *Proceedings of the National Academy of Sciences*, 108(7), 3017–22, http://www.pnas.org/content/pnas/108/7/3017.full.pdf; Erickson et al. (2009), 'Aerobic fitness is associated with hippocampal volume in elderly humans', *Hippocampus*, 19(10), 1030–9, https://www.ncbi.nlm.nih.gov/pmc/articles/PMC3072565/; Thomas et al. (2016), 'Multimodal characterization of rapid anterior hippocampal volume increase associated with aerobic exercise', *Neuroimage*, 131, 162–70, https://www.ncbi.nlm.nih.gov/pmc/articles/PMC4848119/. See also Griffin et al. (2011), 'Aerobic exercise improves hippocampal function and increases BDNF in the serum of young adult males', *Physiology & Behavior*, 104(5), 934–41, https://www.sciencedirect.com/science/article/pii/S0031938411003088, for a similar result in young adults.

**9.** Griffin et al., op. cit.
**10.** This astounding brain-imaging capability brings with it a central problem, though: you must have a theory of what activity occurs where, why, and over what timescale to support a particular function or process in which part or parts of the brain work either in isolation, or more probably, in concert. More than this, you must also have a theory of what order activity might occur in, within and between differing brain regions. This means, in turn, that the design of tasks to try and understand the working brain is crucial. Control experiments will be required: these are conditions in which no change or manipulation is made, so that they allow a baseline against which the changes caused by some manipulation might be assessed. Without adequate controls, you cannot know whether or not the changes you have observed have arisen merely as the result of chance or because of an experimental manipulation. How will you know you haven't obtained what are known as 'false positive' results? Controls, statistical analysis, a good theory, decent experiments, clear thinking, a willingness not to fool yourself, and being able to let go a theory slain by data are all key requirements for brain-imaging experimenters (and scientists generally!).
**11.** Ladouce et al. (2017), 'Understanding minds in real-world environments: toward a mobile cognition approach', *Frontiers in human neuroscience*, 10, 694, https://www.frontiersin.org/articles/10.3389/fnhum.2016.00694/full.
**12.** Experience sampling allow us to understand what people are thinking and feeling during the course of their lives such as when out walking. Csikszentmihalyi, M. and Larson, R. (2014), *Validity and reliability of the experience-sampling method. In Flow and the foundations of positive psychology* (Springer), 35–54.
**13.** Fu et al (2014), 'A cortical circuit for gain control by behavioral state', *Cell*, 156(6), 1139–52, https://www.sciencedirect.com/science/article/pii/S0092867414001445); Dadarlat, M. C. and Stryker, M. P. (2017), 'Locomotion enhances neural encoding of visual stimuli in mouse V1', *Journal of Neuroscience*, 2728–16, http://www.jneurosci.org/content/jneuro/early/2017/03/06/JNEUROSCI.2728-16.2017.full.pdf.
**14.** http://www.iceman.it/en/the-iceman/(Ötzi the Iceman Archaeologic sensation, media star, research topic, museum object); Oeggl et al. (2007), 'The reconstruction of the last itinerary of "Ötzi", the Neolithic Iceman, by pollen analyses from sequentially sampled gut extracts', *Quaternary Science Reviews*, 26(7–8), 853–61; https://s3.amazonaws.com/academia.edu.documents/41301635/The_reconstruction_of_the_last_itinerary20160118-13142-1a3jpae.pdf; Paterlini, M. (2011), 'Anthropology: The Iceman defrosted', *Nature*, 471(7336), 34, https://www.researchgate.net/profile/Marta_Paterlini/publication/50267692_Anthropology_The_Iceman_defrosted/links/58ad462f92851c3cfda0705c/Anthropology-The-Iceman-defrosted.pdf.
**15.** https://www.washingtonpost.com/ … iceman/0d60afe8-a3c6-4a9c-acfa-16c9147b40d4/.

**16.** Ardigò et al. (2011), 'Physiological adaptation of a mature adult walking the Alps', *Wilderness & environmental medicine*, 22(3), 236–41, https://www.wemjournal.org/article/S1080-6032(11)00080-9/fulltext; this in an instructive case of mobile data captured while someone walks a vast distance in the wild over many days. Luca Ardigò and his colleagues at the universities of Verona and Parma studied how a sixty-two-year-old reasonably active male adapted and responded to walking a long trail across the Alps. The unnamed Italian man walked 1,300 km over the Via Alpina.

**17.** Kaplan et al. (2017), 'Coronary atherosclerosis in indigenous South American Tsimane: a cross-sectional cohort study', *The Lancet*, 389(10080), 1730–9, https://www.thelancet.com/journals/lancet/article/PIIS0140-6736(17)30752-3/fulltext?code=lancet-site.

**18.** I often wonder, prompted by this case study and what we know of the effects of exercise on brain and body, if certain types of major depression might be responsive to a long and extended bout of wilderness walking. I know of no evidence that this is so, but it does not seem entirely unreasonable, given, as we shall see, the number of brain and body systems that walking modulates.

**19.** Stone et al. (eds) (1999), *The Science of Self-Report: Implications for Research and Practice* (LEA).

**20.** Althoff et al. (2017), 'Large-scale physical activity data reveal worldwide activity inequality', *Nature*, 547(7663), 336, https://www.ncbi.nlm.nih.gov/pmc/articles/PMC5774986/. Note that in Althoff's study, activity levels are likely to be under-reported, as they captured just pedometer data. Periods spent swimming, playing high-intensity sports such as squash or badminton, or team sports involving some degree of physical contact, will not be captured by the smartphone (although these data are increasingly being captured by other means for professional sportspeople).

## 2: WALKING OUT OF AFRICA

**1.** https://www.chesapeakebay.net/S=0/fieldguide/critter/sea_squirt(Sea Squirt *Molgula manhattensis*); http://tunicate-portal.org/; Corbo et al. (2001), 'The ascidian as a model organism in developmental and evolutionary biology', *Cell*, 106(5), 535–8, https://www.cell.com/fulltext/S0092-8674(01)00481-0; Christiaen et al. (2009), 'The sea squirt Ciona intestinalis', *Cold Spring Harbor Protocols*, 2009(12), pdb-emo138, https://www.researchgate.net/profile/Lionel_Christiaen/publication/41424487_The_Sea_Squirt_Ciona_intestinalis/links/5760590d08ae2b8d20eb5fe7/The-Sea-Squirt-Ciona-intestinalis.pdf.

**2.** Although it does retain some ganglionic or nerve-like functions, partly to control these organs.

**3.** https://www.uas.alaska.edu/arts_sciences/naturalsciences/biology/tamone/catalog/cnidaria/urticina_crassicornis/life_history.htm; Geller et al. (2005), 'Fission in sea anemones: integrative studies of life cycle evolution', *Integrative and Comparative Biology*, 45(4), 615–22, https://academic.oup.com/icb/article/45/4/615/636408.
**4.** https://teara.govt.nz/en/diagram/5355/jellyfish-life-cycle; Katsuki, T., & Greenspan, R. J. (2013), 'Jellyfish nervous systems', *Current Biology*, 23(14), R592–4, https://www.cell.com/current-biology/pdf/S0960-9822(13)00359-X.pdf.
**5.** Lee, R., & Roberts, D. (1997), 'Last interglacial (c.117 kyr) human footprints from South Africa', *South African Journal of Science*, 93(8), 349–50; Roberts, D. L. (2008), 'Last interglacial hominid and associated vertebrate fossil trackways in coastal eolianites, South Africa', *Ichnos*, 15(3–4), 190–207; American Association for the Advancement of Science (1998), 'Humanity's Baby Steps', *Science*, 282(5394), 1635, http://science.sciencemag.org/content/282/5394/1635.1; https://en.wikipedia.org/wiki/Eve%27s_footprint.
**6.** Evolution by natural selection is, unfortunately, a widely misunderstood concept, and has been ever since Darwin first articulated its basic precepts. Evolution by natural selection starts from the observation that all organisms vary in virtually infinite ways – height, weight, longevity, preferred food sources, complexity of nervous system, teeth; the list is virtually endless. This is known as their 'phenotype'. Individual organisms within and between species vary in height, weight, speed of response to the appearance of food sources, or predators, and the like. For evolution by natural selection to work, this variation between organisms must arise from these differences encoded by their genes – there must be a mechanism of inheritance. Thus, the fundamental plan for the body is encoded within an organism's genome. Differences in phenotype come from variations between organisms in their genes; mutations may occur during the passing of genes from one generation to the next, allowing variation between organisms to arise. The environment in turns acts as a kind of a filter for these variations: be slightly taller, you can get food from trees; be slightly smaller, you can scavenge and hunt food sources along the ground; have slightly better vision at night, you might escape a predator. Organisms manifesting the traits that allow them to survive may go on to reproduce; those without these beneficial traits are lost to history, and to time. For evolution by natural selection to work, therefore, it requires enormously long periods of time and involves wasteful numbers of organisms (in fact most species that have ever existed have become extinct). All of these factors added together lead to different rates of survival, and different rates of reproduction, which, over long periods of time, give rise to the traits of the individuals that comprise a genetic population.
**7.** Hardin et al. (2012), *Becker's World of the Cell* (Benjamin Cummings).

**8.** Maniloff, Jack (1996), 'The Minimal Cell Genome: "On Being the Right Size"', *Proceedings of the National Academy of Sciences of the United States of America*, 93(19), 10004–6.

**9.** Woltering et al. (2014), 'Conservation and divergence of regulatory strategies at Hox Loci and the origin of tetrapod digits', *PLoS biology*, 12(1), e1001773, https://journals.plos.org/plosbiology/article?id=10.1371/journal.pbio.1001773. This is a straightforward introduction to hox genes and walking by John Long and Yann Gibert: 'These genes are made for walking another step from fins to limbs', https://theconversation.com/these-genes-are-made-for-walking-another-step-from-fins-to-limbs-22126. Note, however, that the existence of fish that walk along the ocean floor, such as the skate, moderate the stronger claims that a repressor gene inactivated the expression of limbs in these early fish (see next note). The following are more technical introductions to Hox genes and limb segmentation and formation, and the place of the Hox gene in evolution: Shubin et al. (1997), 'Fossils, genes and the evolution of animal limbs', *Nature*, 388(6643), 639, https://www.researchgate.net/publication/13958634_Fossils_genes_and_the_evolution_of_animal_limbs; Burke et al. (1995), 'Hox genes and the evolution of vertebrate axial morphology', *Development*, 121(2), 333–46, http://dev.biologists.org/content/121/2/333.short; Petit, F., Sears, K. E. and Ahituv, N. (2017), 'Limb development: a paradigm of gene regulation', *Nature Reviews Genetics*, 18(4), 245, https://www.researchgate.net/publication/313375342_Limb_development_a_paradigm_of_gene_regulation.

**10.** Lutz et al. (1996), 'Rescue of Drosophila labial null mutant by the chicken ortholog Hoxb-1 demonstrates that the function of Hox genes is phylogenetically conserved', *Genes & development*, 10(2), 176–84, http://genesdev.cshlp.org/content/10/2/176.full.pdf.

**11.** This excellent *National Geographic* article has a wonderful video of the walking skate: https://news.nationalgeographic.com/2018/02/skate-neural-genetics-walking-human-evolution-spd/; see also https://www.sciencedaily.com/releases/2018/02/180208120912.htm; Jung et al. (2018), 'The ancient origins of neural substrates for land walking', *Cell*, 172(4), 667–82, https://www.cell.com/cell/fulltext/S0092-8674(18)30050-3?innerTabvideo-abstract_mmc8=.

**12.** Dawkins, R. (1996), *The Blind Watchmaker: Why the evidence of evolution reveals a universe without design* (W. W. Norton & Company).

**13.** http://www.ucmp.berkeley.edu/vertebrates/tetrapods/tetraintro.html; https://www.sciencedirect.com/topics/veterinary-science-and-veterinary-medicine/tetrapod.

**14.** http://www.valentiaisland.ie/explore-valentia/tetrapod-trackway/; Stössel, I. (1995), 'The discovery of a new Devonian tetrapod trackway in SW Ireland', *Journal* *of the Geological Society*, 152(2), 407–13, https://www.researchgate.net/publication/249545894_The_discovery_of_a_new_Devonian_tetrapod_trackway_in_SW_Ireland. The local authority could do more to support walkers, however. Clearly marked but low-maintenance trails and tracks, with occasional way stations for shelter when the Atlantic weather turns, would be a win for all concerned.

**15.** Apes (https://www2.palomar.edu/anthro/primate/prim_7.htm); Gebo, D. L. (2013), 'Primate Locomotion', *Nature Education Knowledge*, 4(8):1, https://www.nature.com/scitable/knowledge/library/primate-locomotion-105284696; http://www.indiana.edu/~semliki/PDFs/HuntCognitiveDemands.pdf; https://scholar.harvard.edu/files/dlieberman/files/2015f.pdf.

**16.** https://answersafrica.com/african-proverbs-meanings.html; some have disputed the origins of this saying (https://www.npr.org/sections/goatsandsoda/2016/07/30/487925796/it-takes-a-village-to-determine-the-origins-of-an-african-proverb), suggesting that its African origins are unclear, but it does catch an essential truth about walking collectively, as opposed to going it alone.

**17.** Lee, Sang-Hee (2018), 'Where Do We Come From?', *Anthropology News*, 18 September 2018, doi:10.1111/AN.972, http://www.anthropology-news.org/index.php/2018/09/18/where-do-we-come-from/.

**18.** Sankararaman et al. (2012), 'The date of interbreeding between Neanderthals and modern humans', *PLoS genetics*, 8(10), e1002947, https://journals.plos.org/plosgenetics/article?id=10.1371/journal.pgen.1002947.

**19.** http://humanorigins.si.edu/evidence/human-fossils/species/ardipithecus ramidus; White et al. (2009), '*Ardipithecus ramidus* and the paleobiology of early hominids', *Science*, 326(5949), 64–86, http://science.sciencemag.org/content/sci/326/5949/64.full.pdf; Kimbel et al. (2014), '*Ardipithecus ramidus* and the evolution of the human cranial base', *Proceedings of the National Academy of Sciences*, 111(3), 948–53, http://www.pnas.org/content/pnas/111/3/948.full.pdf.

**20.** https://iho.asu.edu/about/lucys-story; the Wikipedia page is a very good introduction: https://en.wikipedia.org/wiki/Lucy_(Australopithecus)

**21.** Gittelman et al. (2015), 'Comprehensive identification and analysis of human accelerated regulatory DNA', *Genome research*, https://genome.cshlp.org/content/25/9/1245; Machnicki et al. (2016), 'First steps of bipedality in hominids: evidence from the atelid and proconsulid pelvis', *PeerJ*, 4, e1521. doi:10.7717/peerj.1521, https://www.ncbi.nlm.nih.gov/pmc/articles/PMC4715437/.

**22.** Bradley et al. (1998), 'Genetics and domestic cattle origins', *Evolutionary Anthropology: Issues, News, and Reviews*, 6, 79–86, https://onlinelibrary.wiley.com/doi/pdf/10.1002/%28SICI%291520-6505%281998%296%3A3%3C79%3A%3AAID-EVAN2%3E3.0.CO%3B2-R); Beja-Pereira et al. (2003), 'Gene-culture coevolution between cattle milk protein genes and human lactase genes', *Nature genetics*, 35(4), 311, https://www.researchgate.net/profile/Andrew_Chamberlain/publication/8993180_Gene-culture_coevolution_between_cattle_milk_protein_genes_and_human_lactase_geneslinks/00b495228d0399e02000000/Gene-culture-coevolution-between-cattle-milk-protein-genes-and-human-lactase-genes.pdf).

**23.** Holowka, N. B. and Lieberman, D. E. (2018), 'Rethinking the evolution of the human foot: insights from experimental research', *Journal of Experimental Biology*, 221(17), jeb17442, https://www.nicholasholowka.com/uploads/2/8/1/2/28124491/holowka_and_lieberman_2018_jeb.pdf.

**24.** Raichlen et al. (2010), 'Laetoli footprints preserve earliest direct evidence of human-like bipedal biomechanics', *PLoS One*, 5(3), e9769, https://journals.plos.org/plosone/article?id=10.1371/journal.pone.0009769. The caveat here, of course, is that it is entirely possible for an earlier variant in the human lineage to have footprints that appear modern in form. These footprints could derive from an Australopithecus, for example.

**25.** Pontzer, H. (2017), 'Economy and endurance in human evolution', *Current Biology*, 27(12), R613–21, https://www.cell.com/current-biology/pdf/S0960-9822(17)30567-5.pdf.

**26.** Pontzer et al. (2012), 'Hunter-gatherer energetics and human obesity', *PLoS One*, 7(7), e40503, https://journals.plos.org/plosone/article?id=10.1371/journal.pone.0040503. They studied thirty Hadza adults, with approximately fifty – fifty representation of males and females who were compared to a similar group of normal weight, drawn from a developed market economy.

**27.** Selinger et al. (2015), 'Humans can continuously optimize energetic cost during walking', *Current Biology*, 25(18), 2452–6, https://www.sciencedirect.com/science/article/pii/S0960982215009586.

**28.** Hall et al. (2019), 'Ultra-processed diets cause excess calorie intake and weight gain: An inpatient randomized controlled trial of ad libitum food intake', Cell Metabolism, https://www.cell.com/cell-metabolism/pdfExtended/S1550-4131(19)30248-7 (see also: https://www.nytimes.com/2019/05/16/well/eat/why-eating-processed-foods-might-make-you-fat.html). This study compares people on an ultra-processed food diet to a relatively unprocessed food diet. Those on the ultra-processed food diet gained approximately 1 kg of extra weight over just fourteen days, whereas those on the non-processed diet lost 1 kg of weight.

**29.** Lieberman, D. E. (2015), 'Is exercise really medicine? An evolutionary perspective', *Current sports medicine reports*, 14(4), 313–19, https://scholar.harvard.edu/dlieberman/publications/exercise-really-medicine-evolutionary-perspective. This is an interesting interview with Lieberman: https://news.harvard.edu/

gazette/story/2018/04/harvard-evolutionary-biologist-daniel-lieberman-on-the-past-present-and-future-of-speed/.

**30.** https://www.nhs.uk/common-health-questions/food-and-diet/what-should-my-daily-intake-of-calories-be/; https://www.cnpp.usda.gov/sites/default/files/usda_food_patterns/EstimatedCalorieNeedsPerDayTable.pdf.

## 3: HOW TO WALK: THE MECHANICS

**1.** Adolph et al. (2012), 'How do you learn to walk? Thousands of steps and dozens of falls per day', *Psychological Science*, 23(11), 1387–94, http://journals.sagepub.com/doi/pdf/10.1177/0956797612446346.

**2.** D'Avella et al. (2003), 'Combinations of muscle synergies in the construction of a natural motor behavior', *Nature Neuroscience*, 6(3), 300, http://e.guigon.free.fr/rsc/article/dAvellaEtAl03.pdf; for the larger debate about the recruitment of muscles during motor control, see Tresch, M. C. and Jarc, A. (2009), 'The case for and against muscle synergies', *Current Opinion in Neurobiology*, 19(6), 601–7, https://www.ncbi.nlm.nih.gov/pmc/articles/PMC2818278/.

**3.** La Fougere et al. (2010), 'Real versus imagined locomotion: a [18F]-FDG PET-fMRI comparison', *Neuroimage*, 50(4), 1589–98, https://s3.amazonaws.com/academia.edu.documents/44014268/Real_versus_imagined_locomotion_a_18F-FD20160322-1395-wrylyz.pdf.

**4.** Berthoz, A. (2000), *The Brain's Sense of Movement* (Harvard University Press); Pozzo et al. (1990), 'Head stabilization during various locomotor tasks in humans', *Experimental Brain Research*, 82(1), 97–106, https://www.researchgate.net/publication/20897989_Head_stabilization_during_various_locomotor_tasks_in_humans_-_I_Normal_subjects; Pozzo et al. (1991), 'Head stabilization during various locomotor tasks in humans', *Experimental Brain Research*, 85(1), 208–17, https://www.researchgate.net/profile/Thierry_Pozzo/publication/21229398_Head_Stabilization_during_Locomotion_Perturbations_Induced_by_Vestibular_Disorders/links/00b4953aa8d3c3714b000000/Head-Stabilization-during-Locomotion-Perturbations-Induced-by-Vestibular-Disorders.pdf.

**5.** Day, B. L. and Fitzpatrick, R. C. (2005), 'The vestibular system', *Current Biology*, 15(15), R583–6, https://www.cell.com/current-biology/pdf/S0960-9822(05)00837-7.pdf; Khan, S. and Chang, R. (2013), 'Anatomy of the vestibular system: a review', *NeuroRehabilitation*, 32(3), 437–43, https://www.researchgate.net/publication/236642391_Anatomy_of_the_vestibular_system_A_review.

**6.** Hobson et al. (1998), 'Sleep and vestibular adaptation: Implications for function in microgravity', *Journal of Vestibular Research*, 8(1), 81–94, https://pdfs.semanticscholar.org/a8bd/5e00ae66990b59b47a07425ea6f536d7a9ef.pdf.

**7.** https://www.awatrees.com/2014/05/18/asleep-with-our-arboreal-ancestors/; https://www.newscientist.com/article/mg21128335-200-anthropologist-i-slept-up-a-tree-to-understand-chimps/; http://www.bbc.com/earth/story/20150415-apes-reveal-sleep-secrets.

**8.** Murray et al. (2018), 'Balance Control Mediated by Vestibular Circuits Directing Limb Extension or Antagonist Muscle Co-activation', *Cell reports*, 22(5), 1325–38, https://www.sciencedirect.com/science/article/pii/S2211124718300263.

**9.** Pandolf, K. B. and Burr, R. E. (2002), *Medical Aspects of Harsh Environments. Volume 2* (Walter Reed Army Medical Center), http://www.dtic.mil/dtic/tr/fulltext/u2/a433963.pdf.

**10.** Cha, Y.H. (2009), 'Mal de débarquement', *Seminars in Neurology*, 29(5), 520; https://www.ncbi.nlm.nih.gov/pmc/articles/PMC2846419/.

**11.** Salinas et al. (2017), 'How humans use visual optic flow to regulate stepping during walking', *Gait & Posture*, 57, 15–20, https://www.researchgate.net/publication/316802386_How_Humans_Use_Visual_Optic_Flow_to_Regulate_Stepping_During_Walking.

**12.** Kuo, A.D. (2007), 'The six determinants of gait and the inverted pendulum analogy: A dynamic walking perspective', *Human Movement Science*, 26, 617–56, http://citeseerx.ist.psu.edu/viewdoc/download?doi=10.1.1.570.8263&rep=rep1&type=pdf.

**13.** Dimitrijevic et al. (1998), 'Evidence for a Spinal Central Pattern Generator in Humans', *Annals of the New York Academy of Sciences*, 860(1), 360–76, https://www.researchgate.net/publication/13361553_Evidence_for_a_spinal_central_pattern_generator_in_humans_Ann_N_Y_Acad_Sci; Duysens, J. and Van de Crommert, H. W. (1998), 'Neural control of locomotion; Part 1: The central pattern generator from cats to humans', *Gait & posture*, 7(2), 131–41, https://repository.ubn.ru.nl//bitstream/handle/2066/24493/24493___.PDF?sequence=1; Calancie et al. (1994), 'Involuntary stepping after chronic spinal cord injury: evidence for a central rhythm generator for locomotion in man', *Brain*, 117(5), 1143–59, http://citeseerx.ist.psu.edu/viewdoc/download?doi=10.1.1.666.9110&rep=rep1&type=pdf.

**14.** The word 'kinaethesis' is also sometimes loosely used to refer to the specific signals arising from muscles and movement.

**15.** Shirai, N. and Imura, T. (2014), 'Looking away before moving forward: Changes in optic-flow perception precede locomotor development', *Psychological science*, 25(2), 485–93, https://www.researchgate.net/publication/259499202_Looking_Away_Before_Moving_Forward.

**16.** Garrett et al. (2002), 'Locomotor milestones and baby walkers: cross sectional study', *BMJ*, 324(7352), 1494, https://www.bmj.com/content/324/7352/1494.full.

## 4: HOW TO WALK: WHERE ARE YOU GOING?

**1.** Mittelstaedt, M. L. and Mittelstaedt, H. (1980), 'Homing by path integration in a mammal', *Naturwissenschaften*, 67(11), 566–7, https://www.researchgate.net/publication/227274996_Homing_by_path_integration_in_a_mammal; Etienne, A. S. and Jeffery, K. J. (2004), 'Path integration in mammals', *Hippocampus*, 14(2), 180–92, https://onlinelibrary.wiley.com/doi/pdf/10.1002/hipo.10173.

**2.** Loomis et al. (1993), 'Nonvisual navigation by blind and sighted: assessment of path integration ability', *Journal of Experimental Psychology: General*, 122(1), 73; https://pdfs.semanticscholar.org/b7d7/10824cd468d1ce420046b66574dfb2f08cd8.pdf. Note, in an experiment of this type it would be easy to load the dice to get the result that you might want. Loomis and his colleagues were exceptionally careful to control for the demographics of the participant groups. All were approximately the same age: they were either employed or were college students, and all were capable of walking normally, eliminating sources of bias and noise that would otherwise confound their experiments.

**3.** http://www.nasonline.org/publications/biographical-memoirs/memoir-pdfs/tolman-edward.pdf; Tolman, E. C. (1948), 'Cognitive maps in rats and men', *Psychological review*, 55(4), 189, https://pdfs.semanticscholar.org/0874/a64d60a23a20303877e23caf8e1d4bb446a4.pdf.

**4.** Holland, P. C. (2008), 'Cognitive versus stimulus-response theories of learning', *Learning & behavior*, 36(3), 227–41, https://link.springer.com/content/pdf/10.3758/LB.36.3.227.pdf; I have, of course, done a little violence to the theoretical subtleties indulged in by certain behaviourists (such as Clark Hull) with this characterisation, but it is close enough to give the general picture. We present a short discussion of Hull's 'Drive Reduction Theory' in the context of modern ideas regarding motivation here: Callaghan et al. (2018), 'Potential roles for opioid receptors in motivation and major depressive disorder', *Progress in Brain Research* 239, 89–119. Hull's work offers more to modern theories of motivation and his work has to some extent been overlooked.

**5.** Koffka, K. (2013), *Principles of Gestalt Psychology* (Routledge). The treatment of gestalt psychology on Wikipedia is comprehensive and is a good starting point, https://en.wikipedia.org/wiki/Gestalt_psychology: 'This principle maintains that when the human mind (perceptual system) forms a percept or "gestalt", the whole has a reality of its own, independent of the parts. The original famous phrase of gestalt psychologist Kurt Koffka, "the whole is *something else* than the sum of its parts", is often incorrectly translated as "The whole is *greater* than the sum of its parts", and thus used when explaining gestalt theory, and further incorrectly applied to systems theory. Koffka did not like the translation. He firmly corrected students who replaced "other" with "greater". "This is not a principle of addition," he said. The whole has an independent existence.'

**6.** Souman et al. (2009), 'Walking straight into circles', *Current biology*, 19(18), 1538–42, https://www.sciencedirect.com/science/article/pii/S0960982209014791.
**7.** Brunec et al. (2017), 'Contracted time and expanded space: The impact of circumnavigation on judgements of space and time', *Cognition*, 166, 425–32, https://www.sciencedirect.com/science/article/pii/S001002771730166X. (This paper also has perhaps one of the greatest lines in a research paper that I have ever read: 'After each delivery, they [the experimental participants] were teleported to the starting point and given a new goal'! A remarkable technological achievement indeed, previously seen only on *Star Trek* ...)
**8.** Corkin, S. (2013), *Permanent Present Tense: The man with no memory, and what he taught the world* (Penguin UK).
**9.** O'Keefe, J. and Burgess, N. (1999), 'Theta activity, virtual navigation and the human hippocampus', *Trends in cognitive sciences*, 3(11), 403–6, http://memory.psych.upenn.edu/files/pubs/KahaEtal99b.pdf.
**10.** Aghajan, Z. M. et al. (2017), 'Theta oscillations in the human medial temporal lobe during real-world ambulatory movement', *Current Biology*, 27(24), 3743–51, https://www.sciencedirect.com/science/article/pii/S0960982217313994.
**11.** O'Keefe, J. and Dostrovsky, J. (1971), 'The hippocampus as a spatial map. Preliminary evidence from unit activity in the freely-moving rat', *Brain Research*, 34(1): 171–5, http://europepmc.org/abstract/MED/5124915; O'Keefe, J. and Nadel, L. (1978), *The Hippocampus as a Cognitive Map* (Clarendon Press), http://www.cognitivemap.net/HCMpdf/HCMComplete.pdf.
**12.** Ekstrom et al. (2003), 'Cellular networks underlying human spatial navigation', *Nature*, 425(6954), 184, https://www.researchgate.net/profile/Arne_Ekstrom/publication/10573102_Cellular_Networks_underlying_human_spatial_navigation/links/5405cdd20cf2c48563b1b5d4.pdf.
**13.** Maguire et al. (1997), 'Recalling routes around London: activation of the right hippocampus in taxi drivers', *Journal of Neuroscience*, 17(18), 7103–10, http://www.jneurosci.org/content/jneuro/17/18/7103.full.pdf; Maguire et al. (1998), 'Knowing where things are: Parahippocampal involvement in encoding object locations in virtual large-scale space', *Journal of Cognitive Neuroscience*, 10(1), 61–76, https://www.mitpressjournals.org/doi/abs/10.1162/089892998563789.
**14.** Ranck Jr, J. B. (1973), 'Studies on single neurons in dorsal hippocampal formation and septum in unrestrained rats: Part I. Behavioral correlates and firing repertoires', *Experimental Neurology*, 41(2), 461–31, https://deepblue.lib.umich.edu/bitstream/handle/2027.42/33782/0000036.pdf?sequence=1&isAllowed=y.

**15.** Taube et al. (1990), 'Head-direction cells recorded from the postsubiculum in freely moving rats. I. Description and quantitative analysis', *Journal of Neuroscience*, 10(2), 420–35, http://www.jneurosci.org/content/jneuro/10/2/420.full.pdf.
**16.** https://www.nobelprize.org/prizes/medicine/2014/press-release/.
**17.** Grieves, R. M. and Jeffery, K. J. (2017), 'The representation of space in the brain', *Behavioural Processes*, 135, 113–31, http://discovery.ucl.ac.uk/1535831/1/Jeffery_Accepted%20version%20for%20OA%20repository.pdf.
**18.** O'Mara, S. M. (2013), 'The anterior thalamus provides a subcortical circuit supporting memory and spatial navigation', *Frontiers in Systems Neuroscience*, 7, 45, https://www.frontiersin.org/articles/10.3389/fnsys.2013.00045/full; Jankowski et al. (2015), 'Evidence for spatially-responsive neurons in the rostral thalamus', *Frontiers in Behavioral Neuroscience*, 9, 256, https://www.frontiersin.org/articles/10.3389/fnbeh.2015.00256/full); Jankowski, M. M. and O'Mara, S. M. (2015), 'Dynamics of place, boundary and object encoding in rat anterior claustrum', *Frontiers in Behavioral Neuroscience*, 9, 250, https://www.frontiersin.org/articles/10.3389/fnbeh.2015.00250/full; Tsanov, M. and O'Mara, S. M. (2015), 'Decoding signal processing in thalamo-hippocampal circuitry: implications for theories of memory and spatial processing', *Brain Research*, 1621, 368–79, http://www.tara.tcd.ie/bitstream/handle/2262/73409/1-s2.0-S0006899314016722-main.pdf?sequence=1.
**19.** O'Mara, S. M. (2013), 'The anterior thalamus provides a subcortical circuit supporting memory and spatial navigation', *Frontiers in systems neuroscience*, 7, 45, https://www.frontiersin.org/articles/10.3389/fnsys.2013.00045/full; Jankowski et al. (2015), 'Evidence for spatially-responsive neurons in the rostral thalamus', *Frontiers in behavioral neuroscience*, 9, 256, https://www.frontiersin.org/articles/10.3389/fnbeh.2015.00256/full; Jankowski et al. (2015), 'Dynamics of place, boundary and object encoding in rat anterior claustrum', *Frontiers in behavioral neuroscience*, 9, 250, https://www.frontiersin.org/articles/10.3389/fnbeh.2015.00250/full; Tsanov, M. and O'Mara, S. (2015), 'Decoding signal processing in thalamo-hippocampal circuitry: implications for theories of memory and spatial processing', Brain Research, 1621, 368–79, http://www.tara.tcd.ie/bitstream/handle/2262/73409/1-s2.0-S0006899314016722-main.pdf?sequence=1.

## 5: WALKING THE CITY

**1.** This is a great piece in a great series about walking the city: https://www.theguardian.com/cities/2018/oct/05/desire-paths-the-illicit-trails-that-defy-the-urban-planners?CMP=share_btn_tw; see https://www.theguardian.com/

cities/series/walking-the-city for the full collection of articles. For desire lines, see: https://www.witpress.com/Secure/elibrary/papers/SC12/SC12003FU1.pdf. For Robert Macfarlane, see: https://twitter.com/RobGMacfarlane/status/977787226133278725.

**2.** https://www.parliament.uk/about/living-heritage/building/palace/architecture/palacestructure/churchill/.

**3.** https://catalog.data.gov/dataset/walkability-index; http://health-design.spph.ubc.ca/tools/walkability-index/.

**4.** Welch, T., 'Eco of Bologna – Fellini's Rimini – and Ferrara', https://www.eurotrib.com/story/2008/3/30/16150/7191; see also Saitta, D., 'Umberto Eco, Planning Education, and Urban Space', https://www.planetizen.com/node/84742/umberto-eco-planning-education-and-urban-space.

**5.** Althoff et al. (2017), 'Large-scale physical activity data reveal worldwide activity inequality', *Nature*, 547(7663), 336, https://www.ncbi.nlm.nih.gov/pmc/articles/PMC5774986/; see also http://activityinequality.stanford.edu/.

**6.** Speck, J. (2013), *Walkable City: How downtown can save America, one step at a time* (Macmillan).

**7.** Fuller, R. A. and Gaston, K. J. (2009), 'The scaling of green space coverage in European cities', *Biology Letters*, 5(3), 352–5, http://rsbl.royalsocietypublishing.org/content/5/3/352.short.

**8.** https://academic.oup.com/ije/article/34/6/1435/707557.

**9.** Asher et al. (2012), 'Most older pedestrians are unable to cross the road in time: a cross-sectional study', *Age and Ageing*, 41(5), 690–4, https://www.researchgate.net/publication/225307198_Most_older_pedestrians_are_unable_to_cross_the_road_in_time_A_cross-sectional_study.

**10.** Sander et al. (2014), 'The challenges of human population ageing', *Age and Ageing*, 44(2), 185–7, https://academic.oup.com/ageing/article/44/2/185/93994.

**11.** John Gay (1716), *Trivia, or The Art of Walking the Streets of London*, https://www.poemhunter.com/i/ebooks/pdf/john_gay_2012_7.pdf.

**12.** https://www.independent.co.uk/property/house-and-home/rise-and-fall-of-the-boot-scraper-2341628.html.

**13.** https://activelivingresearch.org/sites/default/files/BusinessPerformanceWalkableShoppingAreas_Nov2013.pdf; http://uk.businessinsider.com/millennials-forcing-end-suburban-office-parks-2017-2?r=US&IR=T; https://www.gensler.com/design-forecast-2015-the-future-of-workplace; https://www.nreionline.com/office/do-office-tenants-prefer-city-or-suburbs-answer-complicated; http://www.placemakers.com/2017/11/16/places-that-pay-benefits-of-placemaking-v2/; https://www.vox.com/the-goods/2018/10/26/18025000/walkable-city-walk-score-economy.

**14.** Litman, T (2014), 'The Mobility-Productivity Paradox Exploring the Negative Relationships Between Mobility and Economic Productivity', http://www.vtpi.org/ITED_paradox.pdf.
**15.** Bornstein, M. H. and Bornstein, H. G. (1976), 'The pace of life', *Nature*, 259(5544), 557, https://www.nature.com/articles/259557a0.
**16.** Walmsley, D. J. and Lewis, G. J. (1989), 'The pace of pedestrian flows in cities', *Environment and Behavior*, 21(2), 123–50, https://www.researchgate.net/publication/240689640_The_Pace_of_Pedestrian_Flows_in_Cities.
**17.** Wirtz, P. and Ries, G. (1992), 'The pace of life-reanalysed: Why does walking speed of pedestrians correlate with city size?', *Behaviour*, 123(1), 77–83, https://www.jstor.org/stable/4535062?seq=1#metadata_info_tab_contents.
**18.** Levine, R. V. and Norenzayan, A. (1999), 'The pace of life in 31 countries', *Journal of cross-cultural psychology*, 30(2), 178–205, http://journals.sagepub.com/doi/pdf/10.1177/0022022199030002003.
**19.** Shadmehr et al. (2016), 'A representation of effort in decision-making and motor control', *Current Biology*, 26(14), 1929–34, http://reprints.shadmehrlab.org/Shadmehr_CurrBiol_2016.pdf.
**20.** James, L. (2015), 'Managing Walking Rage: Self-Assessment and Self-Change Techniques', *Journal of Psychological Clinical Psychiatry*, 2(1), 00057.
**21.** https://www.researchgate.net/profile/Leon_James/publication/270823144_Managing_Walking_Rage_Self-Assessment_and_Self-Change_Techniques/links/56baa95408ae6a0040ae01b9/Managing-Walking-Rage-Self-Assessment-and-Self-Change-Techniques.pdf.
**22.** Pelphrey et al. (2004), 'When strangers pass: processing of mutual and averted social gaze in the superior temporal sulcus', *Psychological Science*, 15(9), 598–603, http://journals.sagepub.com/doi/pdf/10.1111/j.0956-7976.2004.00726.x.
**23.** Amaral et al. (1983), 'Evidence for a direct projection from the superior temporal gyrus to the entorhinal cortex in the monkey', *Brain research*, 275(2), 263–77.
**24.** Alexander et al (2016), 'Social and novel contexts modify hippocampal CA2 representations of space', *Nature Communications*, 7, 10300.
**25.** Milgram et al. (1969), 'Note on the drawing power of crowds of different size', *Journal of Personality and Social Psychology*, 13(2), 79, https://www.researchgate.net/profile/Leonard_Bickman/publication/232493453_Note_on_the_Drawing_Power_of_Crowds_of_Different_Size/links/0deec52cf116b0afd3000000/Note-on-the-Drawing-Power-of-Crowds-of-Different-Size.pdf.
**26.** Gallup et al. (2012), 'Visual attention and the acquisition of information in human crowds', *Proceedings of the National Academy of Sciences*, 109(19), 7245–50, http://www.pnas.org/content/pnas/109/19/7245.full.pdf.

## 6: A BALM FOR BODY AND BRAIN

**1.** Klepeis et al. (2001), 'The National Human Activity Pattern Survey (NHAPS): a resource for assessing exposure to environmental pollutants', *Journal of Exposure Science and Environmental Epidemiology*, 11(3), 231, https://indoor.lbl.gov/sites/all/files/lbnl-47713.pdf.

**2.** *2018 Physical Activity Guidelines Advisory Committee Scientific Report*, https://health.gov/paguidelines/second-edition/report/; Biswas et al. (2015), 'Sedentary time and its association with risk for disease incidence, mortality, and hospitalization in adults: a systematic review and meta-analysis', *Annals of Internal Medicine*, 162(2), 123–32, doi:10.7326/M14-165.

**3.** Stephan et al. (2018), 'Physical activity and personality development over twenty years: Evidence from three longitudinal samples', *Journal of research in personality*, 73, 173–9, https://www.ncbi.nlm.nih.gov/pmc/articles/PMC5892442/. These findings are essentially epidemiological – in other words there were no experimental manipulations, but they attempt to control for the problems that come with such a study; however, other studies tend to come to similar conclusions – being sedentary causes a negative drift in the five factors of personality (see https://scholar.google.com/scholar?cites=5366658728132198651&as_sdt=2005&sciodt=0,5&hl=en for others).

**4.** Goldberg, L. R. (1990), '"An alternative" description of personality: the big-five factor structure', *Journal of personality and social psychology*, 59(6), 1216, https://cmapspublic2.ihmc.us/rid=1LQBQ96VY-19DH2XW-GW/Goldberg.Big-Five-FactorsStructure.JPSP.1990.pdf.

**5.** Rodrigues, A. D. (2015), 'Beyond contemplation, the real functions held at the cloisters', *Cloister gardens, courtyards and monastic enclosures*, 13, https://www.researchgate.net/profile/Magdalena_Merlos/publication/283638013_Variations_around_one_constant_The_cloister_typology_in_the_cultural_landscape_of_Aranjuez/links/564cf3b508aeafc2aaafaa9e.pdf#page=13.

**6.** Giovanni Boccaccio, *The Decameron. Translated by Guido Waldman with introduction and notes by Jonathan Usher* (Oxford University Press), xxix, 698.

**7.** Nisbet, E. K. and Zelenski, J. M. (2011), 'Underestimating nearby nature: Affective forecasting errors obscure the happy path to sustainability', *Psychological Science*, 22(9), 1101–6, http://journals.sagepub.com/doi/pdf/10.1177/0956797611418527.

**8.** Wilson and Gilbert (2003), 'Affective Forecasting', *Advances in Experimental Social Psychology*, 35, 345–411.

**9.** See this piece for some wonderful photography: https://www.newyorker.com/culture/photo-booth/japanese-photographer-captures-the-mysterious-power-of-forest-bathing; and this for a tourist's guide to forest bathing: https://

savvytokyo.com/shinrin-yoku-the-japanese-art-of-forest-bathing/. It is too early to be able to say what exactly the health benefits of forest bathing are because proper and controlled dose-response studies have not yet been undertaken, e.g. Oh et al. (2017), 'Health and well-being benefits of spending time in forests: systematic review', *Environmental health and preventive medicine*, 22(1), 71, doi:10.1186/s12199-017-0677-9, and, especially, Shanahan et al. (2015), 'The health benefits of urban nature: how much do we need?', *BioScience*, 65(5), 476–85, https://academic.oup.com/bioscience/article/65/5/476/324489.

**10.** Lovelock, J., (1995), *The Ages of Gaia: A Biography of Our Living Earth* (W. W. Norton & Company).

**11.** Thompson et al. (2012), 'More green space is linked to less stress in deprived communities: Evidence from salivary cortisol patterns', *Landscape and urban planning*, 105(3), 221–9, https://www.sciencedirect.com/science/article/pii/S0169204611003665; this study comes with some caveats: it is correlational, rather than causal, and it is a small, exploratory study; the sample size was small; they conducted the study in January, when daylight hours in Scotland are relatively restricted, and the environment has yet to flower; they had no wealthy or higher socio-economic status controls; and nor did they conduct any comparisons between other cities.

**12.** Thompson, C. W. et al. (2012), 'More green space is linked to less stress in deprived communities: Evidence from salivary cortisol patterns', *Landscape and urban planning*, 105(3), 221–9, https://www.mdpi.com/1660–4601/13/4/440/htm.

**13.** Something else may be going on, of course, but in order to understand the relationships we need large-scale studies to be conducted of the size and type needed if a pharma company were seeking to bring a new drug to the market.

**14.** Kaplan, S. (1995), 'The restorative benefits of nature: Toward an integrative framework', *Journal of Environmental Psychology*, 15(3), 169–82, http://willsull.net/resources/KaplanS1995.pdf; Kaplan, R. and Kaplan, S. (1989), *The Experience of Nature: A psychological perspective* (Cambridge University Press).

**15.** White et al. (2013), 'Feelings of restoration from recent nature visits', *Journal of Environmental Psychology*, 35, 40–51, https://www.researchgate.net/publication/273422708_Feelings_of_restoration_from_recent_nature_visits.

**16.** https://www.nimh.nih.gov/health/topics/depression/index.shtml.

**17.** Kessler, R. C. and Bromet, E. J. (2013), 'The epidemiology of depression across cultures', *Annual review of public health*, 34, 119–38, https://www.ncbi.nlm.nih.gov/pmc/articles/PMC4100461/.

**18.** Cooney et al., 'Exercise for depression', *Cochrane Database of Systematic Reviews 2013*, Issue 9, Art. No. CD004366, doi:10.1002/14651858.CD004366.pub6.

**19.** Harvey et al. (2017), 'Exercise and the prevention of depression: results of the HUNT Cohort Study', *American Journal of Psychiatry*, 175(1), 28–36, https://ajp.psychiatryonline.org/doi/pdf/10.1176/appi.ajp.2017.16111223?casa_token=4zbPWwv9LLQAAAAA:v97geAouVEtev5geGCnHWNuBg41Ju-RTsNfdcr3VjTHFZC6nIylfj37wMZZnNxmTDPP9Z_7m7HE3.
**20.** Simon, G. (2017), 'Should psychiatrists write the exercise prescription for depression?', *American Journal of Psychiatry*, 175(1), 2–3, https://ajp.psychiatryonline.org/doi/10.1176/appi.ajp.2017.17090990.
**21.** Therefore, there are many caveats with the literature available, because it is largely observational and correlative. Another approach, of course, is a straightforward experimental one. In this case the approach is to use experimental animal models of depression, and then titrate the dose of exercise and the timing of that dose, relative to the treatment that causes them to have a depressive-like symptomatology. Here, as we will see, the literature offers a much cleaner result. It seems very likely, based on the experimental animal literature, that lots of walking (or aerobic exercise more generally) acts both as an inoculant against depressive-like behaviours, and as a treatment which reduces their severity, at least comparably to the best available pharmacological antidepressants on the market.
**22.** Relatively speaking, few studies focus on the pleasure and reward felt after an intense physical workout. Here's one example: Frazão et al. (2016), 'Feeling of pleasure to high-intensity interval exercise is dependent on the number of work bouts and physical activity status', *PLoS One*, 11(3), e0152752, https://journals.plos.org/plosone/article?id=10.1371/journal.pone.0152752; also see Ekkekakis, P. (2003), 'Pleasure and displeasure from the body: Perspectives from exercise', *Cognition and Emotion*, 17(2), 213–39, https://www.researchgate.net/publication/247496658_Pleasure_and_displeasure_from_the_body_Perspectives_from_exercise, which suggests that the source of pleasure (and the contrary experience of dysphoria) shifts from cognitive to interoceptive sources over the course of exercise.
**23.** https://www.nobelprize.org/prizes/literature/1950/russell/lecture/.
**24.** The literature is so great that just some of the most interesting and consequential specimen results will be described and discussed here. For a sampling, go to: https://scholar.google.com/scholar?hl=en&as_sdt=0%2C5&q=learning+memory+aerobic+exercise&oq=learning+memory+aerobic+exer#d=gs_hdr_drw&p=&u=.
**25.** Hebb, D. O. (1949), *The Organization of Behaviour* (John Wiley), http://s-f-walker.org.uk/pubsebooks/pdfs/The_Organization_of_Behavior-Donald_O._Hebb.pdf.
**26.** Thoenen, H. (1995), 'Neurotrophins and neuronal plasticity', *Science*, 270(5236), 593–8, http://science.sciencemag.org/content/270/5236/593; Leal et al. (2014),

'BDNF-induced local protein synthesis and synaptic plasticity', *Neuropharmacology*, 76, 639–56, https://estudogeral.sib.uc.pt/bitstream/10316/25252/1/1-s2.0-S0028390813001421-main.pdf; de Melo Coelho et al. (2013), 'Physical exercise modulates peripheral levels of brain-derived neurotrophic factor (BDNF): a systematic review of experimental studies in the elderly', *Archives of gerontology and geriatrics*, 56(1), 10–15, https://www.researchgate.net/publication/228101079_Physical_exercise_modulates_peripheral_levels_of_Brain-Derived_Neurotrophic_Factor_BDNF_a_systematic_review_of_experimental_studies_in_the_elderly.

**27.** This often-replicated result was originally made by Carl Cotman's research group (https://www.faculty.uci.edu/profile.cfm?faculty_id=2273) – see, for example, Neeper et al. (1995), 'Exercise and brain neurotrophins', *Nature*, 373, 109; Oliff et al. (1998), 'Exercise-induced regulation of brain-derived neurotrophic factor (BDNF) transcripts in the rat hippocampus', *Molecular Brain Research*, 61(1–2), 147–53; Adlard et al. (2005), 'The exercise-induced expression of BDNF within the hippocampus varies across life-span', *Neurobiology of Aging*, 26(4), 511–20; Cotman et al. (2007), 'Exercise builds brain health: key roles of growth factor cascades and inflammation', *Trends in Neurosciences*, 30(9), 464–72. My research group is one of many to have replicated this result, and we have additionally shown that exercise-induced increases in BDNF effectively abrogate the action of small molecules that otherwise are associated with a pro-inflammatory effect in the brain – see, for example, Shaw et al. (2003), 'Deficits in spatial learning and synaptic plasticity induced by the rapid and competitive broad-spectrum cyclooxygenase inhibitor ibuprofen are reversed by increasing endogenous brain-derived neurotrophic factor', *European Journal of Neuroscience*, 17(11), 2438–46; Callaghan et al. (2017), 'Exercise prevents IFN-α-induced mood and cognitive dysfunction and increases BDNF expression in the rat', *Physiology & Behavior*, 179, 377–83.

**28.** Aggleton et al. (2010), 'Hippocampal–anterior thalamic pathways for memory: uncovering a network of direct and indirect actions', *European Journal of Neuroscience*, 31(12), 2292–307.

**29.** This is a very good summary of findings on the risks of running compared to walking: https://www.vox.com/2015/8/4/9091093/walking-versus-running; see also https://academic.oup.com/ije/article/39/2/580/679411.

**30.** Suter et al (1994), 'Jogging or walking – comparison of health effects,' *Annals of Epidemiology*, 4(5), 375–81, https://www.sciencedirect.com/science/article/pii/1047279794900728.

**31.** Davitt et al. (2018), 'Moderate-vigorous Intensity Run Vs. Walk On Hemodynamics, Metabolism And Perception Of Effort: 1942 Board# 203 May 31 3', *Medicine & Science in Sports & Exercise*, 50(5S), 468–9, https://insights.ovid.com/medicine-

science-sports-exercise/mespex/2018/05/001/moderate-vigorous-intensity-run-vs-walk/1539/00005768.
**32.** Rich et al. (2017), 'Skeletal myofiber vascular endothelial growth factor is required for the exercise training-induced increase in dentate gyrus neuronal precursor cells', *Journal of Physiology*, 595(17), 5931–43, https://physoc.onlinelibrary.wiley.com/doi/pdf/10.1113/JP273994.
**33.** Demangel et al. (2017), 'Early structural and functional signature of 3-day human skeletal muscle disuse using the dry immersion model', *Journal of Physiology*, 595(13), 4301–15, https://physoc.onlinelibrary.wiley.com/doi/pdf/10.1113/JP273895. They gathered a group of twelve male participants (average age thirty-two), and then measured muscle strength, muscle function and took MRI images through the thigh muscles.

## 7: CREATIVE WALKING

**1.** Friedrich Nietzsche, *Twilight of the Idols*, http://www.inp.uw.edu.pl/mdsie/Political_Thought/twilight-of-the-idols-friedrich-neitzsche.pdf. Others (e.g. https://www.goodreads.com/quotes/42472-all-truly-great-thoughts-are-conceived-while-walking) have rendered this quote as 'All truly great thoughts are conceived while walking'.
**2.** Henry David Thoreau, *The Portable Thoreau*, http://www.penguin.com/ajax/books/excerpt/9780143106500; see also http://blogthoreau.blogspot.com/2014/08/a-thousand-rills-thoreaus-journal-19.html.
**3.** Corn, A. (1999), 'The Wordsworth Retrospective', *Hudson Review*, 52(3), 359–78, https://www.jstor.org/stable/3853432?seq=3#metadata_info_tab_ contents.
**4.** https://www.psychologytoday.com/us/blog/the-interrogated-brain/201812/the-importance-daily-rituals-creativity; Currey, M. (ed.) (2013), *Daily Rituals: How Artists Work* (Knopf).
**5.** Russell, B. (1967–9), *The Autobiography of Bertrand Russell*, 3 vols. (Allen & Unwin).
**6.** Orlet, C., (2004), 'The Gymnasiums of the Mind', *Philosophy Now*, https://philosophynow.org/issues/44/The_Gymnasiums_of_the_Mind.
**7.** Raichle et al. (2001), 'A default mode of brain function', *Proceedings of the National Academy of Sciences*, 98(2), 676–82; https://www.pnas.org/content/pnas/98/2/676.full.pdf.
**8.** Christoff et al. (2009), 'Experience sampling during fMRI reveals default network and executive system contributions to mind wandering', *Proceedings of the National Academy of Sciences*, 106(21), 8719–24, https://www.pnas.org/content/pnas/106/21/8719.full.pdf.

**9.** Baird et al. (2012), 'Inspired by distraction: mind wandering facilitates creative incubation', *Psychological Science*, 23(10), 1117–22, https://journals.sagepub.com/doi/pdf/10.1177/0956797612446024.
**10.** Gusnard et al. (2001), 'Medial prefrontal cortex and self-referential mental activity: relation to a default mode of brain function', *Proceedings of the National Academy of Sciences*, 98(7), 4259–64, https://www.pnas.org/content/pnas/106/6/1942.full.pdf; Farb, et al. (2007), 'Attending to the present: mindfulness meditation reveals distinct neural modes of self-reference', *Social cognitive and affective neuroscience*, 2(4), 313–22, https://academic.oup.com/scan/article/2/4/313/1676557.
**11.** Beaty et al. (2018), 'Robust prediction of individual creative ability from brain functional connectivity', *Proceedings of the National Academy of Sciences*, https://www.pnas.org/content/pnas/early/2018/01/09/1713532115.full.pdf.
**12.** Kahneman, D. (2011), *Thinking, Fast and Slow* (Farrar, Straus & Giroux).
**13.** Ibid., 40.
**14.** Peter Lynch, 'The many modern uses of quaternions: A surprising application is to electric toothbrushes but they have many vital functions', https://www.irishtimes.com/news/science/the-many-modern-uses-of-quaternions-1.3642385#.W7YHhykoyeY.twitter.
**15.** https://math.berkeley.edu/~robin/Hamilton/fourth.html.
**16.** Runco, M. A. and Jaeger, G. J. (2012), 'The standard definition of creativity', *Creativity Research Journal*, 21, 92–6.
**17.** Olufsen et al. (2005), 'Blood pressure and blood flow variation during postural change from sitting to standing: model development and validation', *Journal of Applied Physiology*, 99(4), 1523–37; Ouchi et al. (1999), 'Brain activation during maintenance of standing postures in humans', *Brain*, 122(2), 329–38.
**18.** Oppezzo, M. and Schwartz, D. L. (2014), 'Give your ideas some legs: The positive effect of walking on creative thinking', *Journal of experimental psychology: learning, memory, and cognition*, 40(4), 1142–52, https://lagunita.stanford.edu/c4x/Medicine/ANES204/asset/Give_Your_Ideas_Some_Legs_2.pdf.
**19.** Steinberg et al. (1997), 'Exercise enhances creativity independently of mood', *British Journal of Sports Medicine*, 31(3), 240–5, https://bjsm.bmj.com/content/bjsports/31/3/240.full.pdf.
**20.** Ning Hao et al. (2017), 'Enhancing creativity: Proper body posture meets proper emotion', *Acta Psychologica*, 173, 32–40.
**21.** https://www.psychologytoday.com/us/blog/our-innovating-minds/201808/does-open-office-plan-make-creative-environment; https://www.economist.com/business/2018/07/28/open-offices-can-lead-to-closed-minds.
**22.** Kiefer et al. (2009), 'Walking changes the dynamics of cognitive estimates of time intervals', *Journal of Experimental Psychology: Human Perception*

*and Performance*, 35(5), 1532, https://www.researchgate.net/profile/Adam_Kiefer/publication/26869873_Walking_Changes_the_Dynamics_of_Cognitive_Estimates_of_Time_Intervals/links/0046351d424175e8fc000000/Walking-Changes-the-Dynamics-of-Cognitive-Estimates-of-Time-Intervals.pdf.

**23.** See Csikszentmihalyi, M. (2014), 'Toward a psychology of optimal experience', *Flow and the foundations of positive psychology* (Springer), 209–26; Csikszentmihalyi, M. (2008), *Flow: The psychology of Optimal Experience* (Harper Perennial Modern Classics); Csikszentmihalyi and LeFevre (1989), 'Optimal experience in work and leisure', *Journal of personality and social psychology*, 56(5), 815, http://citeseerx.ist.psu.edu/viewdoc/download?doi=10.1.1.845.9235&rep=rep1&type=pdf.

**24.** Keinänen, M. (2016), 'Taking your mind for a walk: a qualitative investigation of walking and thinking among nine Norwegian academics', *Higher Education*, 71(4), 593–605, https://link.springer.com/article/10.1007/s10734-015-9926-2.

**25.** Steinbeck, J., *Sweet Thursday*, https://www.penguinrandomhouse.com/books/354543/sweet-thursday-by-john-steinbeck/9780143039471/.

**26.** See, for example, https://web.chemdoodle.com/kekules-dream/ for a short account, and http://acshist.scs.illinois.edu/bulletin_open_access/v31-1/v31-1%20p28-30.pdf for a longer account.

**27.** Wagner et al. (2004), 'Sleep inspires insight', *Nature*, 427(6972), 352, http://www.cogsci.ucsd.edu/~chiba/SleepInsightWagnerNature04.pdf. Ed Yong provides a nice summary of some current thinking on sleep and creativity here: https://www.theatlantic.com/science/archive/2018/05/sleep-creativity-theory/560399/; the emphasis being that good-quality sleep, with all sleep cycles engaged and spent, is key.

## 8: SOCIAL WALKING

**1.** Wong, K. (2011), 'Fossil footprints of early modern humans found in Tanzania', *Scientific American Blog*, https://blogs.scientificamerican.com/observations/fossil-footprints-of-early-modern-humans-found-in-tanzania/; Greshko, M. (2016), 'Treasure Trove of Ancient Human Footprints Found Near Volcano Hundreds of crisscrossing tracks offer a glimpse of life in Africa around 19,000 years ago', https://news.nationalgeographic.com/2016/10/ancient-human-footprints-africa-volcano-science/; Greshko, M. (2018) 'Treasure Trove of Fossil Human Footprints Is Vanishing', https://www.nationalgeographic.com/science/2018/08/news-engare-sero-ol-doinyo-lengai-tanzania-behavior/.

**2.** Twain, M., 'A Tramp Abroad', https://www.gutenberg.org/files/119/119-h/119-h.htm, chapter XXIII.

**3.** https://tilda.tcd.ie/publications/reports/pdf/Report_PhysicalActivity.pdf.
**4.** Clearfield, M. W. (2011), 'Learning to walk changes infants' social interactions', *Infant Behavior and Development*, 34(1), 15–25, https://www.whitman.edu/Documents/Academics/Psychology/Clearfield%202010.pdf.
**5.** Muscatell et al. (2012), 'Social status modulates neural activity in the mentalizing network' *NeuroImage*, 60(3), 1771–77.
**6.** Bonini, L. (2017), 'The extended mirror neuron network: Anatomy, origin, and functions', *The Neuroscientist*, 23(1), 56–67.
**7.** Dunbar, R. (2010), *How many friends does one person need? Dunbar's number and other evolutionary quirks* (Faber & Faber).
**8.** Dunbar, R. (1998), *Grooming, Gossip, and the Evolution of Language* (Harvard University Press).
**9.** Yun et al. (2012), 'Interpersonal body and neural synchronization as a marker of implicit social interaction', *Scientific reports*, 2, 959, https://www.nature.com/articles/srep00959; they recruited twenty healthy right-handed male participants, and equipped them with EEG electrodes. Participants in their study sat across a table from a leader, and were instructed to use either their left or right arm (removing hand preference from analyses of the brain mechanisms responsible for synchronisation).
**10.** Ikeda et al. (2017), 'Steady beat sound facilitates both coordinated group walking and inter-subject neural synchrony', *Frontiers in Human Neuroscience*, 11, 147, https://www.ncbi.nlm.nih.gov/pmc/articles/PMC5366316/. The data from these studies can be difficult to interpret, because localising the signal obtained to particular brain regions can be tricky. Additionally, the data are collected while the person is walking, so they are up and about, and the data are acquired on the fly.
**11.** The effect observed was peculiar to blood flow in the frontopolar regions, because simultaneously conducted measures of blood flow in the skin were completely uncorrelated with either behavioural flow or stepping more generally. This sort of control allows you to rule out non-specific arousal effects as the cause of the effect.
**12.** Van Schaik et al. (2017), 'Measuring mimicry: general corticospinal facilitation during observation of naturalistic behaviour', *European Journal of Neuroscience*, 46(2), 1828–36, https://www.researchgate.net/profile/Johanna_Van_Schaik/publication/317606742_Measuring_Mimicry_General_Corticospinal_Facilitation_During_Observation_of_Naturalistic_Behaviour/links/595622d7aca272fbb37d150d/Measuring-Mimicry-General-Corticospinal-Facilitation-During-Observation-of-Naturalistic-Behaviour.pdf). Van Schaik and colleagues recruited eighteen female participants in their study, measuring the TMS-motor-evoked potential in the right hands of their participants.

**13.** Sweeny et al. (2013), 'Sensitive perception of a person's direction of walking by 4-year-old children', *Developmental Psychology*, 49(11), 2120, https://www.ncbi.nlm.nih.gov/pmc/articles/PMC4305363/.
**14.** Gabriel et al. (2017), 'The psychological importance of collective assembly: Development and validation of the Tendency for Effervescent Assembly Measure (TEAM)', *Psychological Assessment*, 29(11), 1349–62, doi:10.1037/pas0000434, https://www.researchgate.net/publication/314271695_The_Psychological_Importance_of_Collective_Assembly_Development_and_Validation_of_the_Tendency_for_Effervescent_Assembly_Measure_TEAM.
**15.** https://malachiodoherty.com/2008/08/08/lord-bew-on-burntollet/.
**16.** When attempting policy change, we must also focus on how we reason and make decisions. Underlying much of the reasoning we engage in during the course of our lives is a tendency to retrofit arguments and data to a position that we already have taken. This is known formally as *confirmation bias*, and is a widespread and ubiquitous shortcut in human reasoning.

## ACKNOWLEDGEMENTS

**1.** There are multiple approaches being tried, from targeted spinal-cord electrical stimulation – e.g. Wagner et al (2018), 'Targeted neurotechnology restores walking in humans with spinal cord injury', *Nature*, 563(7729), 65, https://www.nature.com/articles/s41586-018-0649-2 – to stem cell transplantation: Raisman, G. (2001), 'Olfactory ensheathing cells – another miracle cure for spinal cord injury?', *Nature Reviews Neuroscience*, 2(5), 369; Tabakow (2014), 'Functional regeneration of supraspinal connections in a patient with transected spinal cord following transplantation of bulbar olfactory ensheathing cells with peripheral nerve bridging', *Cell Transplantation*, 23(12), 1631–55; to the development of brain–machine interfaces: Donati et al. (2016), 'Long-term training with a brain-machine interface-based gait protocol induces partial neurological recovery in paraplegic patients', *Scientific Reports*, 6, 30383, https://www.nature.com/articles/srep30383.

# INDEX